© 2024 Esther Gonstalla

Originally published as *Atlas eines bedrohten Planeten: 155 geniale Grafiken für alle, die die Welt retten wollen*
© 2023 oekom verlag, Waltherstrasse 29, 80337 München, Germany

Library of Congress Control Number: 2024935670

All Island Press books are printed on environmentally responsible materials.

Manufactured in the United States of America
10 9 8 7 6 5 4 3 2 1

Keywords: agriculture, anthroposphere, atmosphere, biosphere, carbon cycle, carbon dioxide, climate change, climate solutions, deforestation, Earth, ecosystems, energy, extinction, fresh water, glaciers, green energy, greenhouse effect, greenhouse gases, hydrosphere, infographics, infrastructure, IPCC, Intergovernmental Panel on Climate Change, oceans, plastic pollution, pollution, solar power, transportation, water cycle, water scarcity, wind power

ATLAS OF A THREATENED PLANET

150 Infographics to Help Anyone Save the World

ESTHER GONSTALLA

ISLANDPRESS | Washington | Covelo

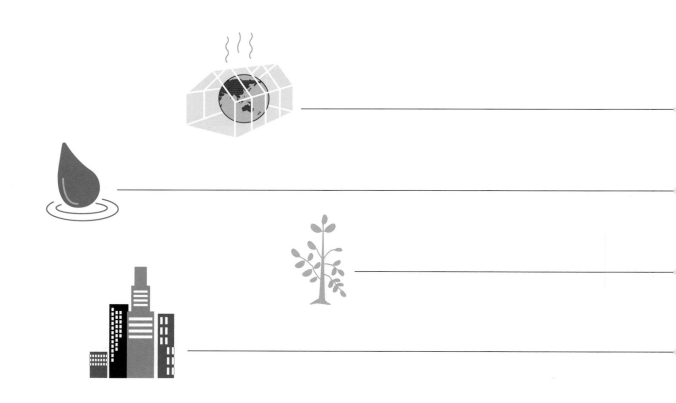

The Spheres of the Earth

Earth's climate
Our climate is controlled by the composition of the atmosphere and is dependent on water vapor content and greenhouse gases.

Oxygen
Animals and humans need oxygen to survive.

The protective layer that surrounds our planet

Weather
Rain, sunshine, clouds, storms: the weather in the atmosphere is constantly changing.

ATMOSPHERE

ANTHROPOSPHERE

Everything man-made and constructed

Agriculture
Livestock farming and cultivation of fruit, vegetables, and energy crops.

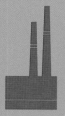

Industry
Production of goods, energy, and waste.

Raw Materials
Production of raw materials such as mining for coal, iron, or rare earth metals, oil production, or clear-cutting for timber.

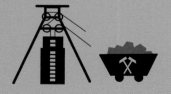

Settlements
From villages to megacities.

Infrastructure
Roads, train tracks, airports, and sea ports for global transport.

Saltwater
The salty oceans make up the majority of our water.

Freshwater
From groundwater to springs, rivers, and lakes to our taps.

Frozen water
Ice, glaciers, snow, frozen soil, and bodies of water.

Life-giving water in all its forms

HYDROSPHERE

Water cycle
Water is constantly exchanged between land and atmosphere.

BIOSPHERE

All animals and plants on, above, and below the ground

Plants
From grasses, cacti, and trees to algae and kelp forests underwater, plants are as diverse as their habitats.

Animals
From tiny sea creatures to the largest mammals on land and migratory birds in the air, every animal has its niche.

Ecosystems
Plants and animals live in complex interconnected habitats.

THE ATMOSPHERE

Air, Weather & Climate

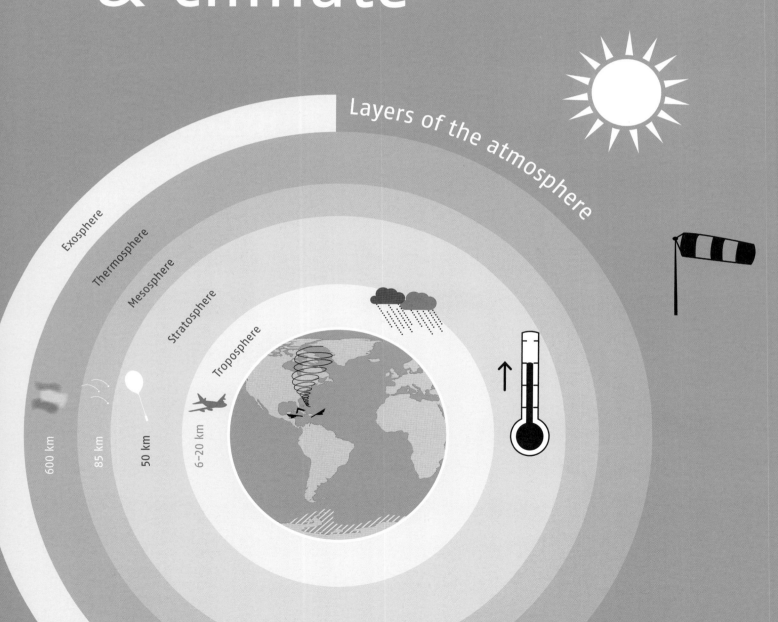

Layers of the atmosphere

Exosphere
Thermosphere
Mesosphere
Stratosphere
Troposphere

600 km
85 km
50 km
6–20 km

At | mo | sphere
[ancient Greek *atmós* = haze
& *sphaîra* = sphere (Earth)]
The atmosphere is the gaseous envelope
of the Earth, inside which wind, clouds
and precipitation form. It filters, absorbs,
and reflects the sun's radiation, making
life on Earth possible. Without an
atmosphere, our planet would resemble
a snowball.

Water Cycle of the Earth

About 70% of the Earth's surface is covered by water. The atmosphere, groundwater, rivers, oceans, and ice masses constantly exchange water. In the process, its state also changes (from liquid to gas, for example).

Water vapor forms clouds...

...many of which are transported to land masses by winds.

Snow

During evaporation, water becomes water vapor. When the sun heats a water surface, the water molecules accelerate until they are so fast that they escape into the air as a gas.

Precipitation

Evaporation

Sublimation

Glaciers & Ice Sheets

Permafrost

Oceans

Rivers & Lakes

Earth mass

Water mass

Compared to the mass of the Earth, the mass of water on our planet is tiny.

Clouds, which form from the rising water vapor, provide rain replenishment.

About **70%** of the precipitation in a forest returns to the atmosphere.

Some water evaporates directly from the canopy.

Water absorbed by the roots is partly transpired through the crown.

Precipitation

Interception

Transpiration

Rain that falls on leaves and needles is first stored in the forest canopy.

Only when more rain falls than the canopy can absorb does the water seep into the forest floor.

Soil evaporation

Deep seepage and filtration

Root uptake

Seepage water

Soil is the smallest source of water evaporation.

Groundwater

Sources: Markart & Kohl (2009), Perlmann et al. (2019), USGS (2019), Zimmermann et al. (2008)

Air in Motion

The air in the atmosphere never stands completely still. This is mainly due to temperature differences between the Earth's surface and the higher layers of air, which create high- and low-pressure areas.

When the sun warms the Earth's surface and thus also the air, water evaporates from the soil and water bodies and enriches the air with moisture; in addition, the air expands as it warms. This creates a low-pressure area: the warm, humid air rises, cools down, and condenses, forming clouds. This is why it is often rainy and stormy in low-pressure areas.

A little further away, the cooled air sinks again and warms up the closer it gets to the ground. This creates an area of high pressure, where sinking air masses cause the clouds to dissipate. This is why high-pressure areas bring bright sunshine.

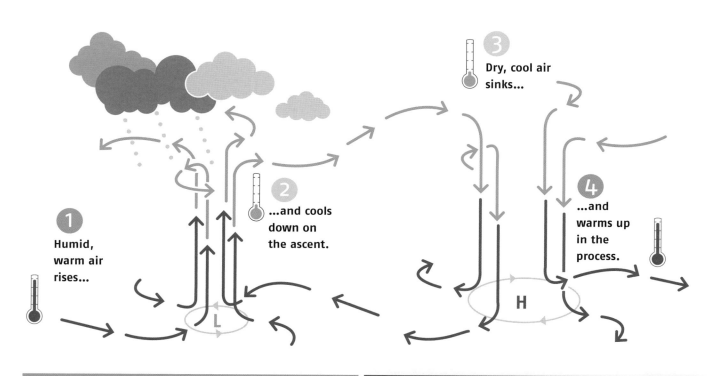

1 Humid, warm air rises...

2 ...and cools down on the ascent.

3 Dry, cool air sinks...

4 ...and warms up in the process.

LOW-PRESSURE AREA	HIGH-PRESSURE AREA

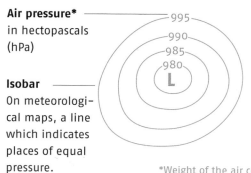

Air pressure*
in hectopascals (hPa)

Isobar
On meteorological maps, a line which indicates places of equal pressure.

995
990
985
980
L

Wind direction:
Counterclockwise from the Earth's center (in the Northern Hemisphere).

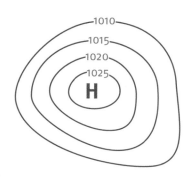

1010
1015
1020
1025
H

Wind direction:
Clockwise from the Earth's center (in the Northern Hemisphere).

*Weight of the air column from the outermost edge of the atmosphere to the Earth's surface.

Source: DWD (2022)

Air Currents at the Poles

Stable polar vortex
In winter, the sun no longer rises at the poles, leading to a strong temperature differential between the cold poles and warm equator. This creates the polar vortex, a fast air current in the stratosphere (an altitude of 16–48 km) that moves in a circle around the north and south poles.

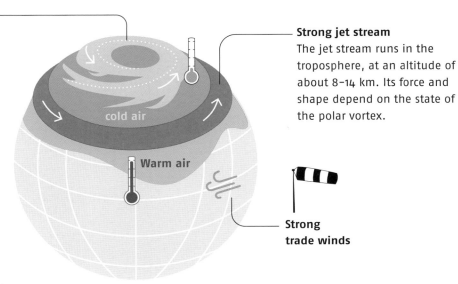

Strong jet stream
The jet stream runs in the troposphere, at an altitude of about 8–14 km. Its force and shape depend on the state of the polar vortex.

cold air

Warm air

Strong trade winds

Disrupted polar vortex
The Arctic polar vortex weakens and breaks into several parts each April (or earlier in exceptional cases such as 2018, 2019, and 2021). This leads to a weaker jet stream and longer-lasting, stationary weather patterns that can become extreme weather events.

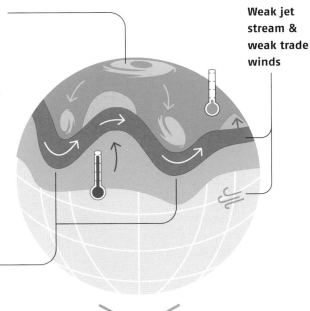

Weak jet stream & weak trade winds

Rossby waves
When the temperature gradient between the pole and the equator becomes very large, the jet stream creates waves, so-called Rossby waves. These can become so strong that they weaken and even split the polar vortex.

Climate change
Scientists disagree on whether and to what extent global warming has an impact on the polar vortex and jet stream. If Rossby waves become stronger due to global warming, it could lead to more severe droughts and a collapse of global food production in regions where about 1/4 of the world's food is grown.

Sources: Coumou et al. (2018), Kornhuber et al. (2019), NOAA (2019, 2021), Voosen (2020), WOR (2019)

Extreme Storms

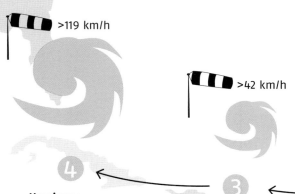

Hurricanes

>119 km/h

>42 km/h

4

Hurricane
Wind speeds of at least 119 km/h are called hurricanes.

3

Tropical cyclone
If the wind speed is at least 42 km/h, it is called a tropical storm.

The movement of the air can take on extreme forms, especially when cyclones form over the water. Depending on the region of the world, we speak of hurricanes (Atlantic and Eastern Pacific), extratropical cyclones (North Atlantic), typhoons (Northwest Pacific), or cyclones (Indian Ocean and Southwest Pacific).

In late summer, when the ocean is at its warmest and evaporation at its highest, hurricane season begins in the Atlantic. If the water temperature at the surface is higher than 26 degrees Celsius, tropical storms develop from low-pressure areas. Global warming may cause this temperature limit to be exceeded more often in the future, and destructive storms such as hurricanes are likely to increase in frequency and strength around the world. This would affect Caribbean and Pacific island nations and heavily populated coastal and river delta areas in North America and Asia, in particular.

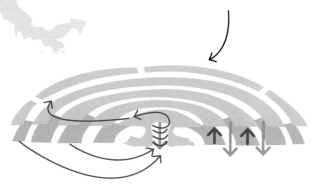

In a hurricane, warm air rises in a spiral while cold air sinks at the center or eye. The more warm humid air rises, the stronger and faster the whirlwind moves.

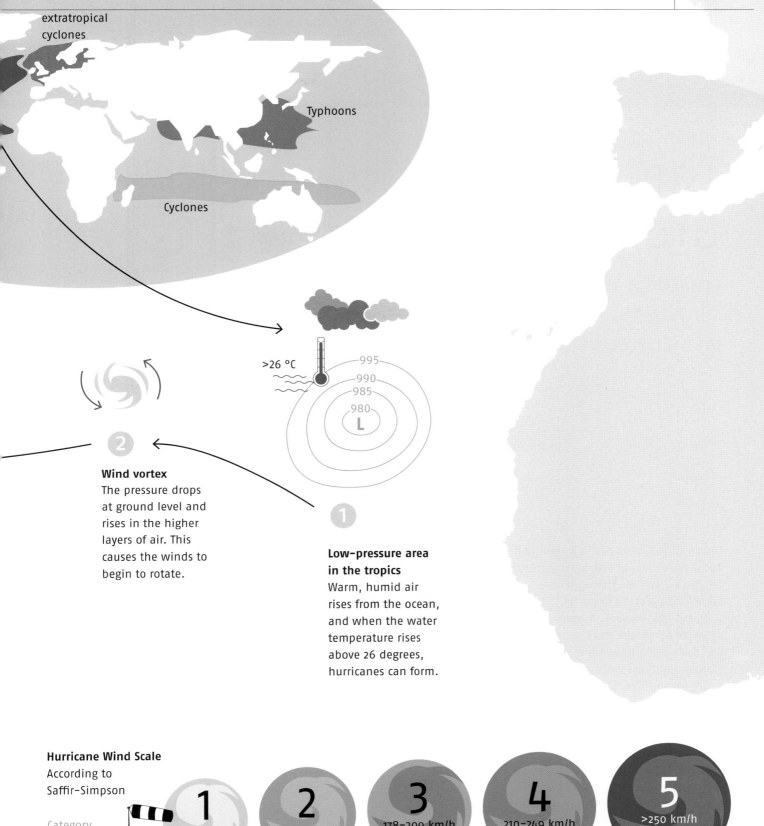

extratropical cyclones

Typhoons

Cyclones

>26 °C

995
990
985
980
L

2

Wind vortex
The pressure drops
at ground level and
rises in the higher
layers of air. This
causes the winds to
begin to rotate.

1

**Low-pressure area
in the tropics**
Warm, humid air
rises from the ocean,
and when the water
temperature rises
above 26 degrees,
hurricanes can form.

Hurricane Wind Scale
According to
Saffir-Simpson

Category
Wind Speeds

Damage

	1 119–153 km/h	2 154–177 km/h	3 178–209 km/h	4 210–249 km/h	5 >250 km/h
	MINIMAL	MODERATE	EXCESSIVE	EXTREME	CATASTROPHIC

Sources: NOAA (2022), Rahmstorf (2018), Velden et al. (2017)

Climate

The Natural Greenhouse Effect

created a moderate, stable climate from the end of the last ice age until about 1880.

Life on Earth is only possible because it has temperatures at which living things can thrive. We have the so-called greenhouse gases to thank for these temperatures. These gases ensure that, when heat from the sun reaches Earth and radiates back toward space, some of it does not disappear, but remains in our atmosphere. Natural greenhouse gases are primarily water vapor, carbon dioxide, and trace gases such as methane.

As long as the concentration of greenhouse gases is in equilibrium, the average temperature on Earth remains almost constant, but when man-made greenhouse gases are added, it becomes warmer and warmer.

More on man-made greenhouse gases starting on p. 20.

Composition of greenhouse gases

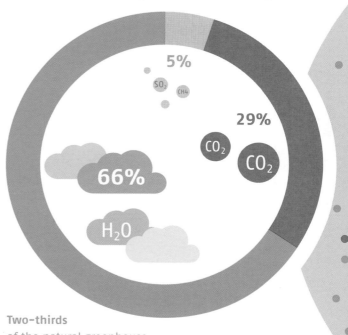

5%
SO_2 CH_4

29%
CO_2 CO_2

66%
H_2O

Two-thirds
of the natural greenhouse effect is caused by water vapor (H_2O), just under a third by CO_2, and a small percentage by other trace gases such as methane (CH_4).

1 dot stands for 1%:
- Water vapor
- Carbon dioxide
- Trace gases

Sources: DWD (2018), Rahmstorf (2013), Riedel & Janiak (2015), UBA (2021), UNFCCC (2022)

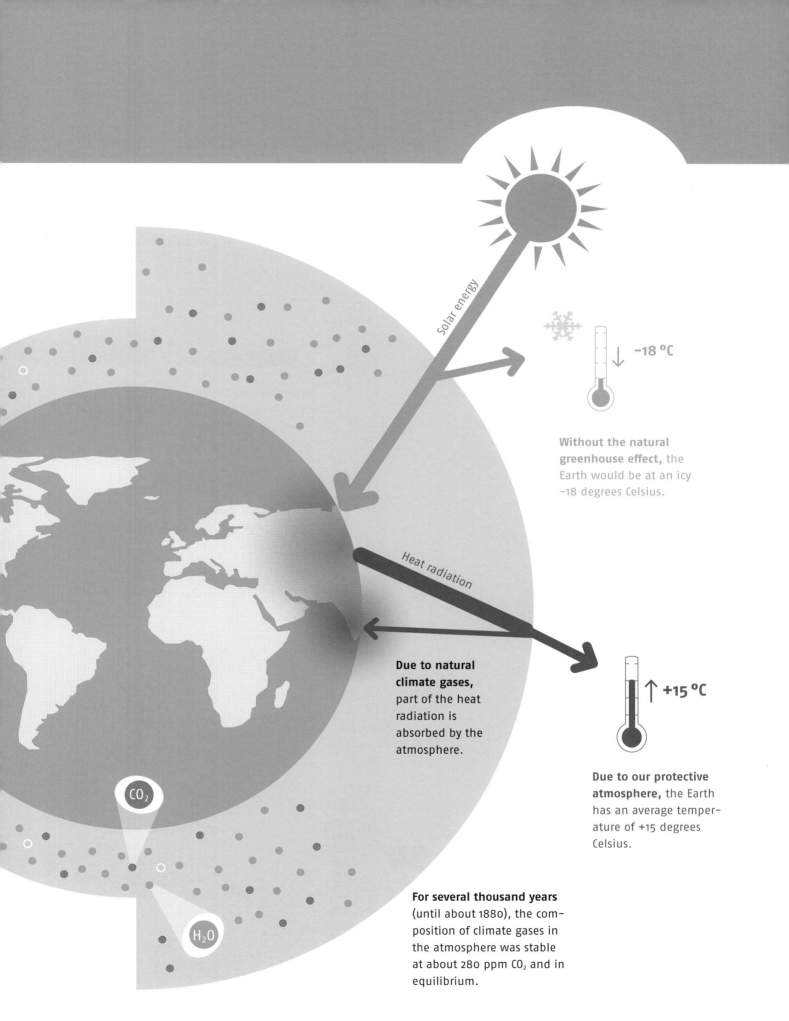

Solar energy

Heat radiation

↓ −18 °C

Without the natural greenhouse effect, the Earth would be at an icy −18 degrees Celsius.

Due to natural climate gases, part of the heat radiation is absorbed by the atmosphere.

↑ +15 °C

Due to our protective atmosphere, the Earth has an average temperature of +15 degrees Celsius.

CO_2

H_2O

For several thousand years (until about 1880), the composition of climate gases in the atmosphere was stable at about 280 ppm CO_2 and in equilibrium.

We Are Changing the Climate

with high CO_2 emissions caused by:

Industrial agriculture, feed production, and factory farming, as well as soil degradation

Burning fossil fuels such as coal, oil, and gas to meet rising global energy demand

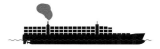

Passenger and freight transport on roads, rivers, seas, and in the air

Forest fires, forest use, native forest logging, and forest management

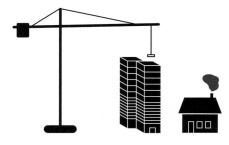

Housing and house construction, heating, and energy demand in the private sector

Production of goods, means of transport, textiles, and furniture in energy-intensive processes

Since the beginning of the industrial revolution (around 1880), humans have been releasing more and more greenhouse gases into the atmosphere, causing the temperature to rise steadily by 1.1 to 1.2 degrees Celsius since 1880. The problem was recognized at the first World Climate Summit in Berlin in 1995. In 1997, the United Nations (UN) Kyoto Protocol was established, committing industrialized countries to reduce their greenhouse gas emissions.

Rising global (online) consumption and lifestyle of abundant disposable goods

PARIS **CLIMATE** SUMMIT 2015

Nearly 200 countries have signed the Paris Climate Agreement. At the 2015 meeting, world leaders agreed to limit global warming to well below 2 degrees Celsius, to ensure that life on our planet remains worth living for today's children and their children.

Anthropogenic Greenhouse Effect
(2021)

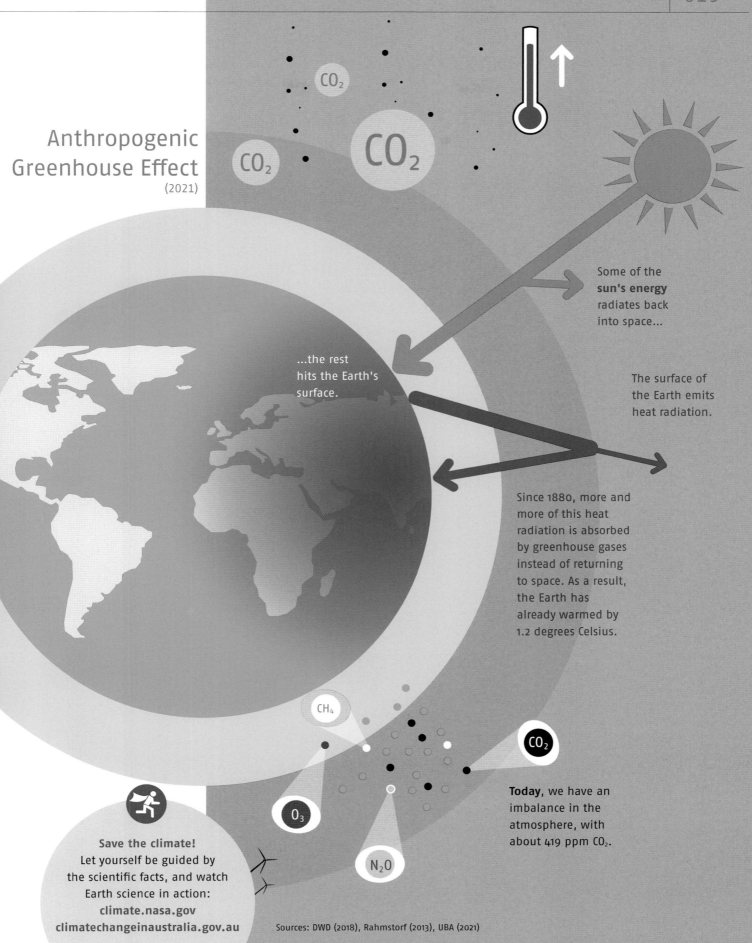

CO_2

CO_2

CO_2

Some of the **sun's energy** radiates back into space...

...the rest hits the Earth's surface.

The surface of the Earth emits heat radiation.

Since 1880, more and more of this heat radiation is absorbed by greenhouse gases instead of returning to space. As a result, the Earth has already warmed by 1.2 degrees Celsius.

CH_4

CO_2

O_3

Today, we have an imbalance in the atmosphere, with about 419 ppm CO_2.

N_2O

Save the climate!
Let yourself be guided by the scientific facts, and watch Earth science in action:
climate.nasa.gov
climatechangeinaustralia.gov.au

Sources: DWD (2018), Rahmstorf (2013), UBA (2021)

Anthropogenic Greenhouse Gases

Scientific evidence gathered over the last 30+ years leaves no doubt that global warming is caused by rising anthropogenic greenhouse gas emissions, particularly the increase in CO_2 levels.

Except for the fluorinated gases, all trace gases shown here were also present in the atmosphere in lower concentrations before the industrial age, but their concentration in the atmosphere is steadily increasing.

64% Carbon dioxide (CO_2)
from fossil fuels and industry

The extraction, processing, and burning of fossil fuels such as coal, natural gas, and oil account for the largest portion of carbon dioxide (CO_2) emissions.

11% Carbon dioxide
from forestry and land use

Deforestation, clearing of forests by fire, draining of peat bogs, and alteration of soils by agriculture also release CO_2 into the atmosphere.

18% Methane

Methane emissions are caused by factory farming, landfills, sewage treatment plants, mining, and extraction of fuels as well as thawing of permafrost.

4% Nitrous oxide (laughing gas)

The majority of nitrous oxide emissions come from agriculture through soil cultivation, nitrogenous fertilizers such as nitrate and ammonia, and factory farming.

2% F-gases (hydrofluorocarbons)

Fluorinated gases are used as propellants, coolants, and extinguishing agents, or in the production of soundproof windows.

Sources: IPCC (WG-III) (2022), UBA (2021)

Greenhouse gas emissions in detail (2019)

Approx. **2/3** water vapor

Water vapor is abundant in our atmosphere. It contributes to two-thirds of the natural greenhouse effect, making it the most important greenhouse gas. As the temperature of the ocean and air continues to rise, the amount of water vapor in the atmosphere also increases. More water vapor leads to more clouds and rain, creating a continuous cycle that has a strong impact on the climate.

Complex Carbon Cycles

Carbon dioxide (CO_2) is the most commonly discussed greenhouse gas when it comes to climate change. Carbon is a fundamental building block of all life on Earth, and it is absorbed, stored, and released back into the atmosphere in a global cycle. Carbon is an essential element for plant growth, as plants use it for photosynthesis to create sugar. When carbon combines with other elements, such as oxygen, nitrogen, or phosphorus, it can lead to the formation of various essential compounds. Plants absorb CO_2 during photosynthesis, but they also release it back into the atmosphere, for example, when leaves decompose.

Human activities such as the burning of oil, gas, and wood contribute significantly to the release of carbon dioxide into the atmosphere. Additionally, human-induced activities like deforestation and land use change have led to large carbon dioxide emissions by destroying ecosystems that have stored a lot of CO_2 over millenia, such as peatlands.

The increasing warming of the oceans due to human-induced climate change has also led to "feedbacks" in the climate system: as the oceans warm, they can absorb less CO_2, causing the concentration of CO_2 in the atmosphere to rise further.

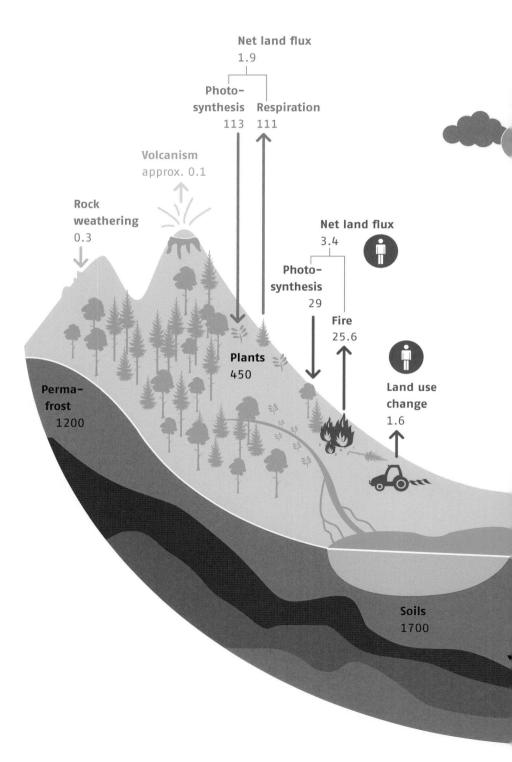

Net land flux
1.9
Photo-synthesis 113   Respiration 111
Volcanism approx. 0.1
Rock weathering 0.3
Net land flux 3.4
Photo-synthesis 29
Fire 25.6
Plants 450
Perma-frost 1200
Land use change 1.6
Soils 1700

Source: IPCC (2021)

OUR "CO₂ BUDGET"

The more CO_2 we emit, the more global average temperature rises. The goal should be a limit of +1.5 degrees Celsius.

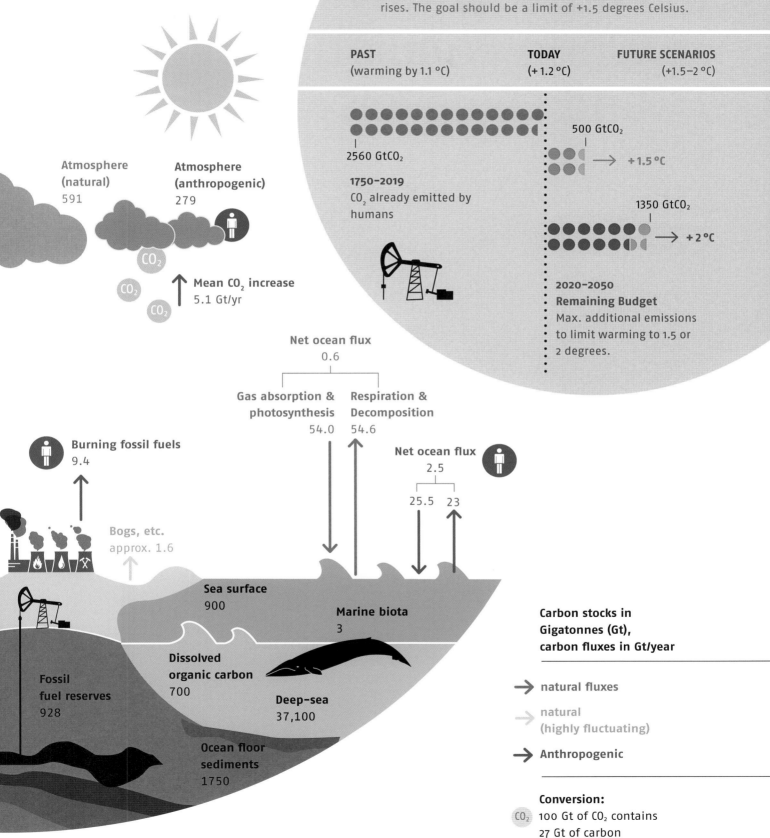

PAST (warming by 1.1 °C)	**TODAY** (+ 1.2 °C)	**FUTURE SCENARIOS** (+1.5–2 °C)

2560 GtCO₂

1750–2019
CO_2 already emitted by humans

500 GtCO₂
→ +1.5 °C

1350 GtCO₂
→ +2 °C

**2020–2050
Remaining Budget**
Max. additional emissions to limit warming to 1.5 or 2 degrees.

Atmosphere (natural)
591

Atmosphere (anthropogenic)
279

CO_2

CO_2

CO_2

Mean CO_2 increase
5.1 Gt/yr

Net ocean flux
0.6

Gas absorption & photosynthesis
54.0

Respiration & Decomposition
54.6

Burning fossil fuels
9.4

Net ocean flux
2.5

25.5   23

Bogs, etc.
approx. 1.6

Sea surface
900

Marine biota
3

Dissolved organic carbon
700

Deep-sea
37,100

Fossil fuel reserves
928

Ocean floor sediments
1750

Carbon stocks in Gigatonnes (Gt), carbon fluxes in Gt/year

→ natural fluxes

→ natural (highly fluctuating)

→ Anthropogenic

Conversion:
CO_2 100 Gt of CO_2 contains 27 Gt of carbon

Parallel Growth: World Population and CO$_2$ levels

7

6

5

4

3

2

World population in billions

1

0

The demand for energy, transportation, and food is increasing as rapidly as the global population, resulting in a rise in CO$_2$ emissions worldwide.

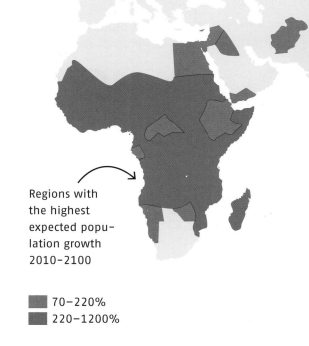

Regions with the highest expected population growth 2010–2100

■ 70–220%
■ 220–1200%

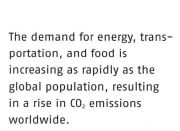

World population 1800:
0.9 billion people
CO$_2$ content: 270 ppm

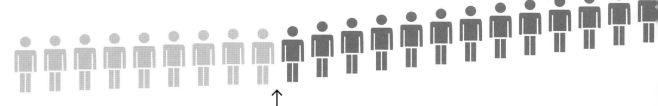

1800

1850

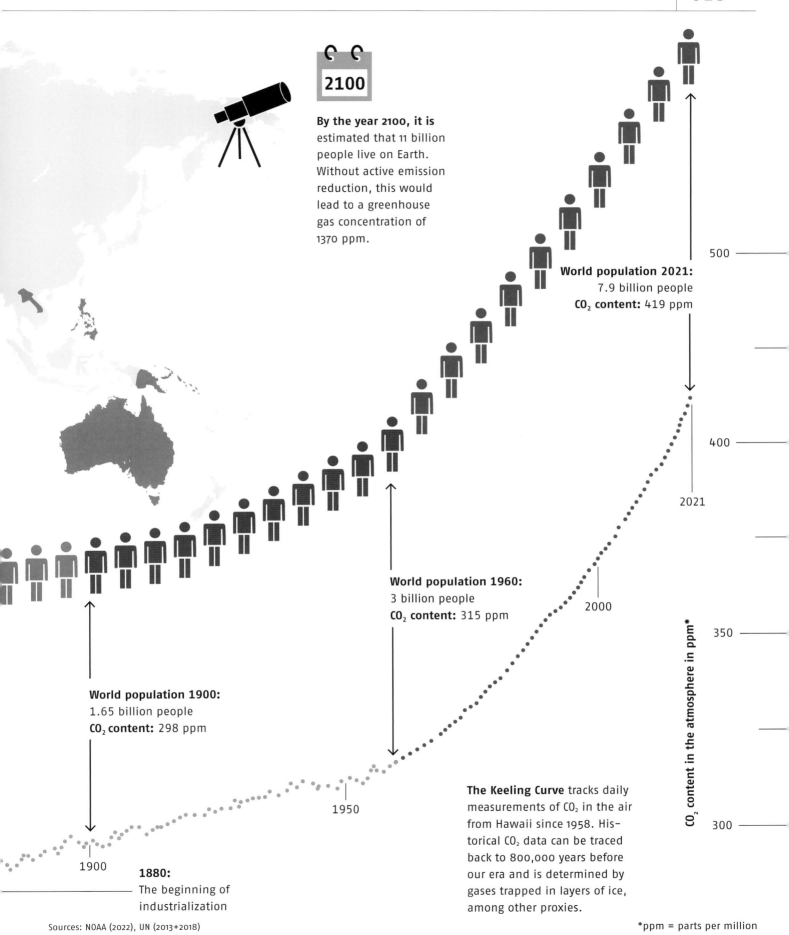

2100

By the year 2100, it is estimated that 11 billion people live on Earth. Without active emission reduction, this would lead to a greenhouse gas concentration of 1370 ppm.

World population 2021:
7.9 billion people
CO_2 content: 419 ppm

500

400

2021

World population 1960:
3 billion people
CO_2 content: 315 ppm

2000

350

World population 1900:
1.65 billion people
CO_2 content: 298 ppm

CO_2 content in the atmosphere in ppm*

1950

The Keeling Curve tracks daily measurements of CO_2 in the air from Hawaii since 1958. Historical CO_2 data can be traced back to 800,000 years before our era and is determined by gases trapped in layers of ice, among other proxies.

300

1900

1880:
The beginning of industrialization

Sources: NOAA (2022), UN (2013+2018)

*ppm = parts per million

CO$_2$ Emissions: Top 10

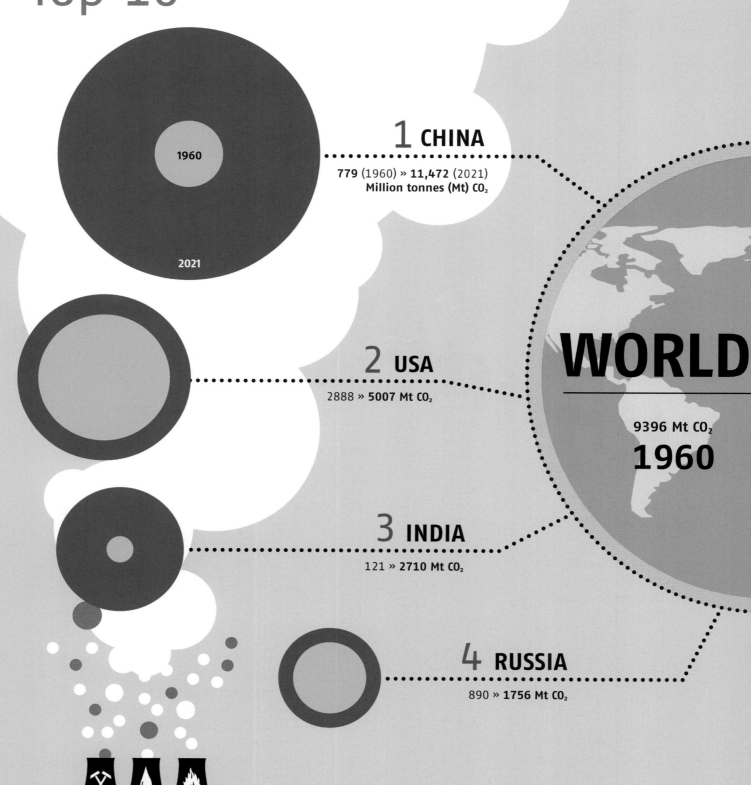

1 CHINA

779 (1960) » **11,472** (2021)
Million tonnes (Mt) CO$_2$

2 USA

2888 » **5007 Mt CO$_2$**

3 INDIA

121 » **2710 Mt CO$_2$**

4 RUSSIA

890 » **1756 Mt CO$_2$**

WORLD

9396 Mt CO$_2$
1960

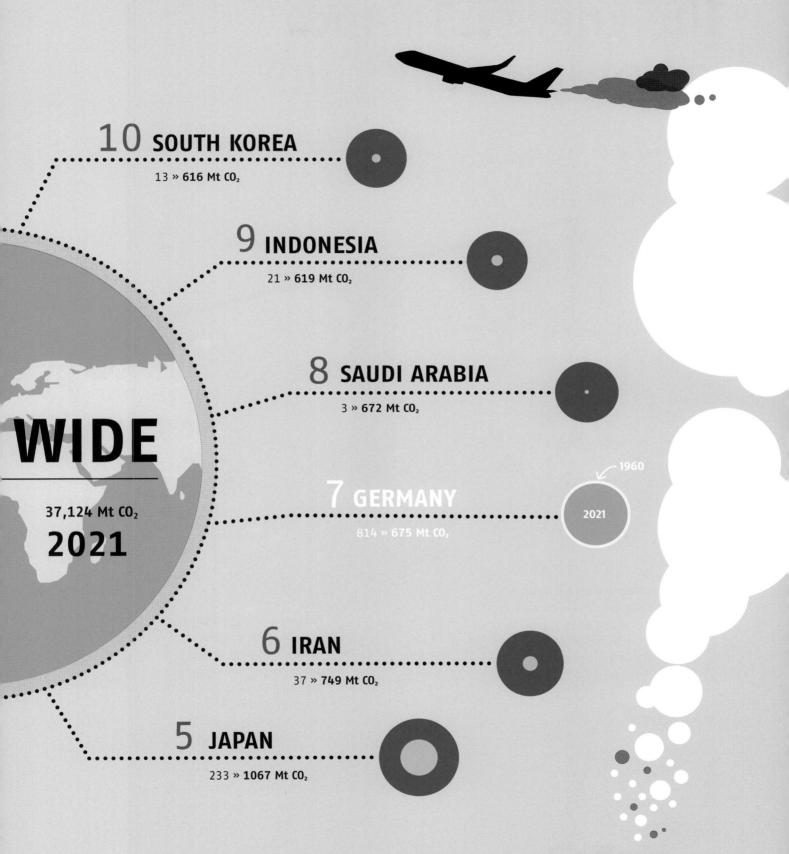

Emissions increase since 1960

Emissions reduction

10 SOUTH KOREA
13 » 616 Mt CO_2

9 INDONESIA
21 » 619 Mt CO_2

8 SAUDI ARABIA
3 » 672 Mt CO_2

7 GERMANY
814 » 675 Mt CO_2

1960

2021

6 IRAN
37 » 749 Mt CO_2

5 JAPAN
233 » 1067 Mt CO_2

WIDE

37,124 Mt CO_2
2021

Source: GCA (2022)

The Energy Balance Is Not in Equilibrium...

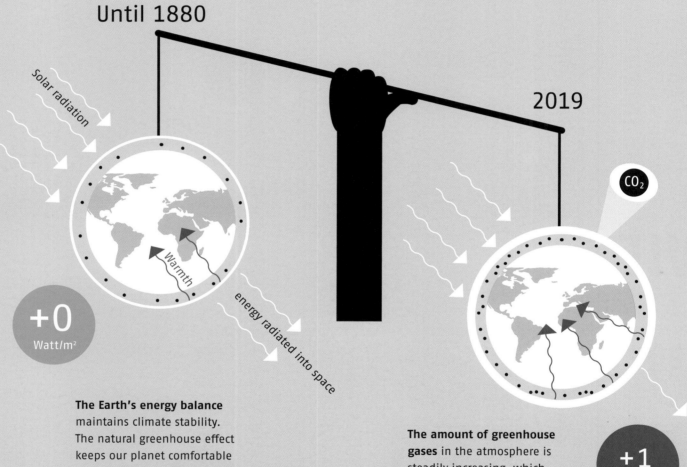

Until 1880

Solar radiation

Warmth

energy radiated into space

+0
Watt/m²

2019

CO₂

+1
Watt/m²

The Earth's energy balance maintains climate stability. The natural greenhouse effect keeps our planet comfortable at an average temperature of 15 degrees Celsius.

The amount of greenhouse gases in the atmosphere is steadily increasing, which means that more heat radiation is being directed back to Earth. Moreover, there is less heat radiation reflected into space due to the dwindling sea ice. As a result, our planet is now warming by about 1 watt per square meter, which is double the figure recorded in 2005.

Sources: Loeb et al. (2021), NASA (2022), NOAA (2022)

...and The Earth Is Warming Up

Temperature change
(worldwide, 1850–2021)

+1.2 °C
warmer

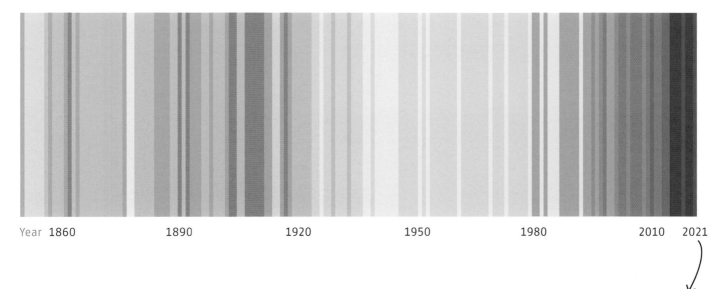

Year 1860   1890   1920   1950   1980   2010   2021

Temperature change in 2021
(compared to 1951–1980)

−4.0   −2.0   −1.0   −0.5   −0.2   0   0.2   0.5   1.0   2.0   **+4.0 °C**

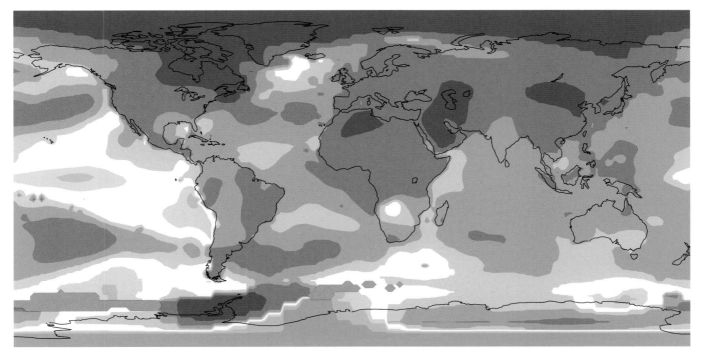

The Distribution of Heat

The additional heat energy generated by the human–induced greenhouse effect is absorbed by:

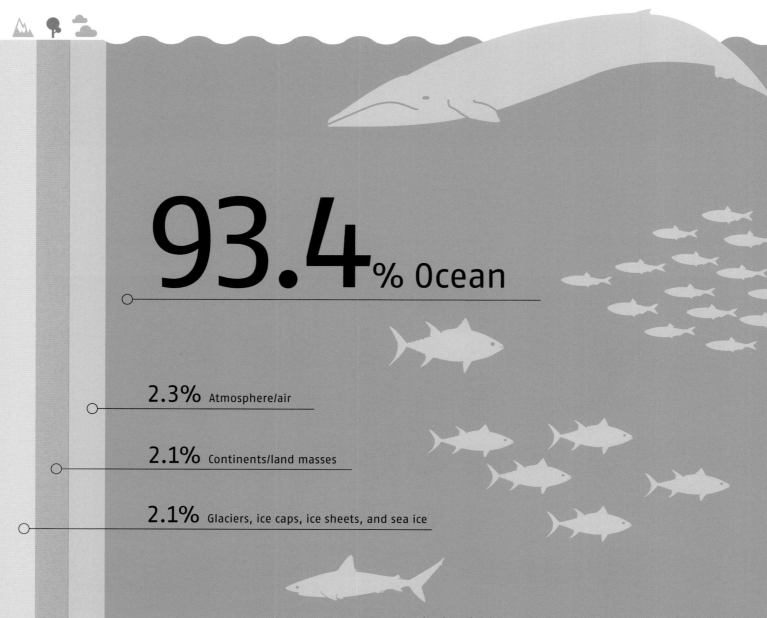

93.4% Ocean

2.3% Atmosphere/air

2.1% Continents/land masses

2.1% Glaciers, ice caps, ice sheets, and sea ice

Sources: Cheng et al. (2019), EPA (2016), Gleckler et al. (2016), IPCC (2021), Resplandy et al. (2018)

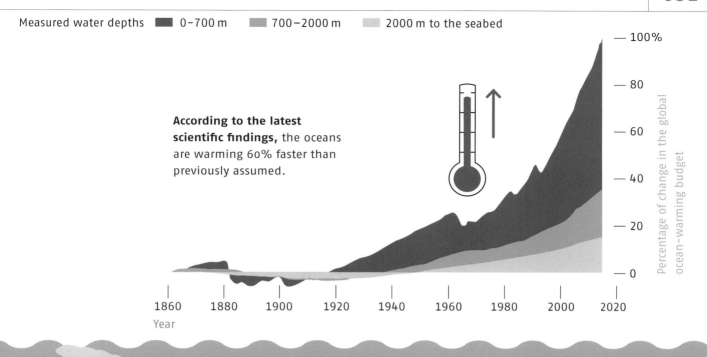

Measured water depths ▪ 0–700 m ▪ 700–2000 m ▪ 2000 m to the seabed

According to the latest scientific findings, the oceans are warming 60% faster than previously assumed.

Percentage of change in the global ocean-warming budget

— 100%
— 80
— 60
— 40
— 20
— 0

1860 1880 1900 1920 1940 1960 1980 2000 2020
Year

Save the seas!
The oceans are important and must be protected from exploitation. Help out and get involved with a local or global NGO, such as Oceana.

Global warming is causing an increase in the temperature of seawater, which in the long term, results in a rise in sea levels. The expansion of the warming liquid and the melting of glaciers and ice sheets contribute to this phenomenon. For instance, if all the ice masses in Greenland were to melt, the global sea level would go up by at least 7 meters.

Researchers predict that by 2100, sea levels will have risen by 80–150 centimeters, including some melting of the ice sheets. However, the melting of ice is difficult to calculate accurately as the dynamics of ice masses change with the warming of oceans. Therefore, it is still uncertain how quickly the Antarctic ice sheets will lose ice and to what extent that will affect sea levels.

see p. 82 and 83

Our Climate Is Already Tipping

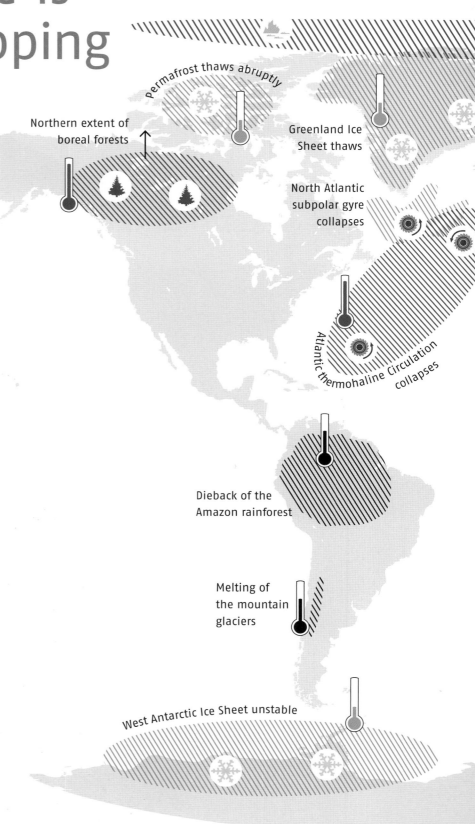

Permafrost thaws abruptly

Northern extent of boreal forests

Greenland Ice Sheet thaws

North Atlantic subpolar gyre collapses

Atlantic thermohaline Circulation collapses

Dieback of the Amazon rainforest

Melting of the mountain glaciers

West Antarctic Ice Sheet unstable

As Earth's temperature warms, not all changes are steady or reversible. Researchers have identified many climate tipping points, critical thresholds that lead to large and accelerating changes when crossed. Recent scientific findings indicate that the climate system may have already reached some tipping points due to the current warming of 1.2 degrees Celsius. Additionally, six other tipping points will become irreversibly triggered with a temperature increase of 1.5 to 2 degrees. The West Antarctic Ice Sheet, which is already unstable, is one such tipping point. Its melting would dangerously increase sea levels. Permafrost in Canada and Russia is already thawing, releasing methane into the atmosphere, which is a potent climate gas. These changes are more threatening because they're likely irreversible.

When the temperature reaches 2 degrees of warming, it may be impossible to stop man-made climate change from intensifying. The consequences of warming will reinforce each other in a domino effect, triggering new life-threatening tipping points.

It is now more urgent than ever to act, as we are heading for 2.6 degrees of warming with the current emission reductions, which is unacceptable.

Source: Armstrong McKay et al. (2022)

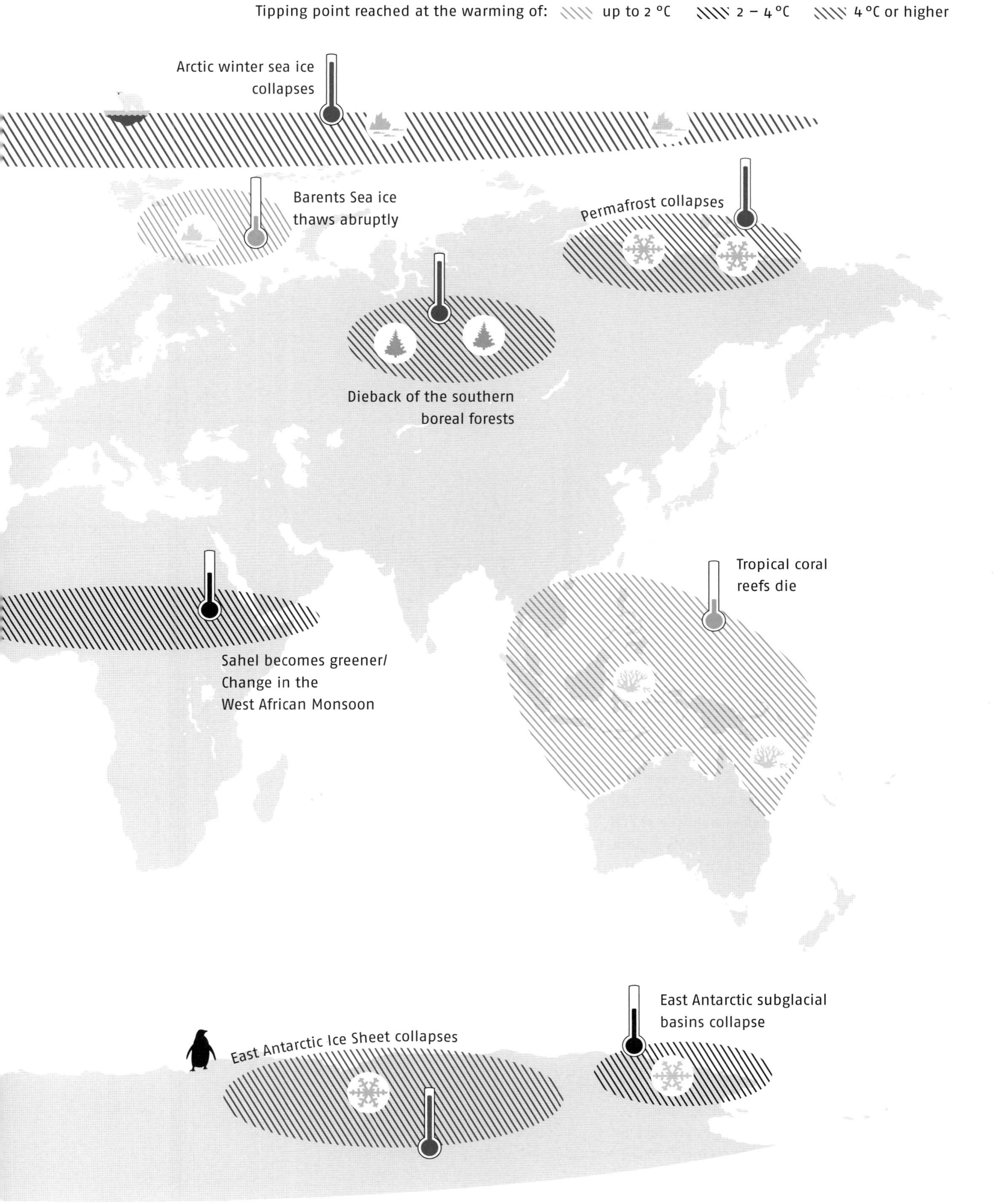

Tipping point reached at the warming of: ///// up to 2 °C ///// 2 – 4 °C ///// 4 °C or higher

Arctic winter sea ice collapses

Barents Sea ice thaws abruptly

Permafrost collapses

Dieback of the southern boreal forests

Tropical coral reefs die

Sahel becomes greener/ Change in the West African Monsoon

East Antarctic subglacial basins collapse

East Antarctic Ice Sheet collapses

Effects of the Climate Crisis

Changes in ecosystems lead, among other things, to declining biodiversity and higher rates of extinction and endangered species.

Increasing water scarcity due to prolonged droughts and higher future demand.

4 °C

By 2100, the African continent is expected to experience a warming of 4°C.

Although Africa is one of the lowest contributors to greenhouse gas emissions, the climate crisis is hitting there particularly hard. The equatorial countries are facing a higher-than-average rise in temperature, and they lack the financial resources to protect themselves from the effects of climate change. One-third of Africa's population is already experiencing prolonged droughts, the drying up of lakes, and difficult agricultural conditions. While some regions are experiencing less rainfall, others, such as West Africa, will face heavier rains, storms, and floods in the future. The climate crisis will exacerbate global shortages of drinking water and food.

The UN predicts that approximately 50 million "environmental refugees" will be forced to leave almost uninhabitable rural areas due to climate change and migrate to larger cities. For every additional degree of warming, approximately 1 billion people worldwide may have to relocate from flood or drought regions.

Sea level rise and extreme weather lead to problems in infrastructure, drinking water supply, and health problems due to flooding, among other things.

Save lives!
Support projects that sustainably green up degraded landscapes and empower farmers, such as Farmer-Managed Natural Regeneration: fmnrhub.com.au.

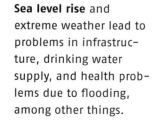

Degradation of coral reefs leads to declining fish stocks and loss of income and food for traditional fishermen.

High crop losses due to droughts and floods lead to food shortages and health problems.

Weakened livestock are more susceptible to disease and premature death due to heat, drinking water shortages, and lack of feed.

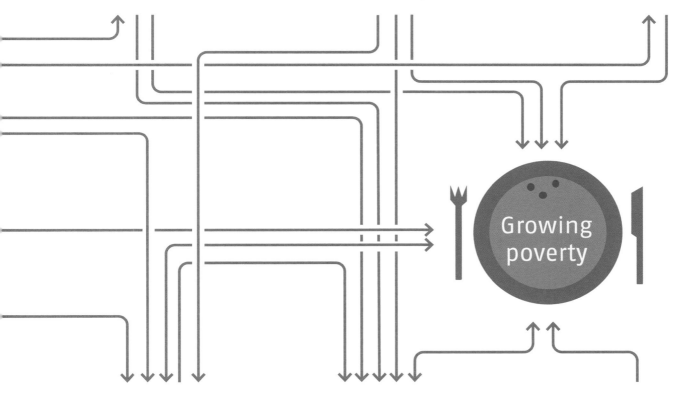

Growing poverty

Climate migration is increasing and can lead to unrest, intensification of violent conflicts, human rights violations, and political instability.

Malnutrition is on the rise in Africa and leads to lifelong health problems. Triggers include climate-related crop failures or water scarcity.

Diseases such as malaria and cholera, which are transmitted by mosquitoes or contaminated water, can become more prevalent or widespread due to heavy rainfall and high temperatures.

Sources: Cho (2021), IPCC (2022)

Extreme Weather

 Snow/ice   Heavy rain   Tropical storms

Warmer winters
Extreme cold is up to 100 times less likely than in 1880.

Extreme drought in Western Canada: 2015
Drought is becoming more severe and likely.

The oceans are warming
High ocean temperatures are more extreme and widespread.

Flooding in Miami since 1994
Flood disasters are 500% more likely.

Extreme hurricane season: 2014
Hurricanes are getting stronger and more frequent.

Drought in the southern Amazon: 2010
Droughts are becoming more prolonged and more likely.

Heatwave in Argentina: 2013
Risk of severe heat is 5 times higher.

The impact of the climate crisis has become apparent worldwide through the increase of extreme weather events.

Since 2000, a separate field of research called "extreme event attribution" has emerged, which involves calculating the likelihood of a weather event being linked to the man-made climate crisis. This research aims to determine the extent to which climate change has influenced thermal processes that may have resulted in the event. This map displays the results of nearly 100 studies where the man-made climate crisis was identified as a trigger or amplifier of extreme weather events.

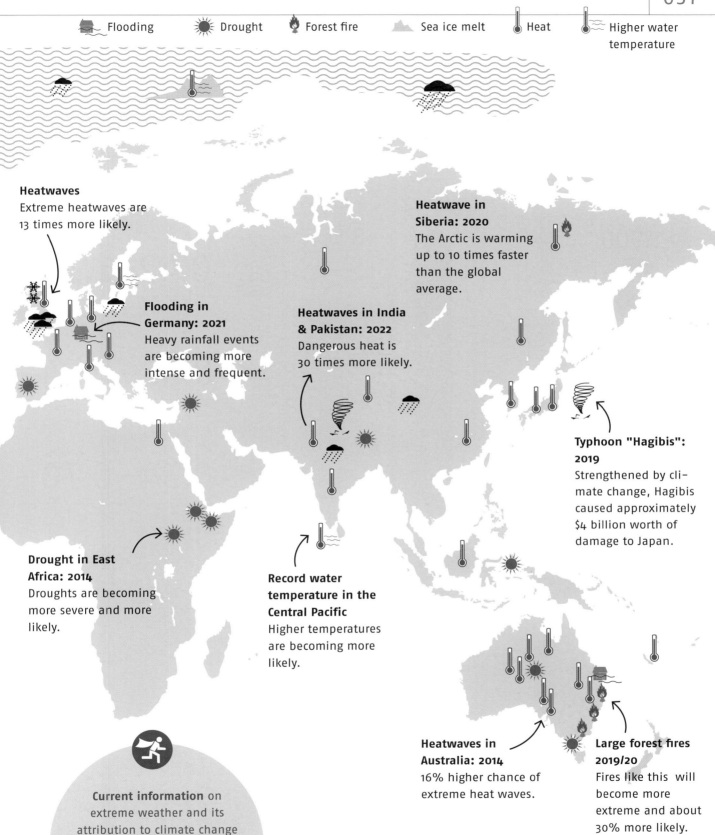

Flooding ● **Drought** ● **Forest fire** ● **Sea ice melt** ● **Heat** ● **Higher water temperature**

Heatwaves
Extreme heatwaves are 13 times more likely.

Heatwave in Siberia: 2020
The Arctic is warming up to 10 times faster than the global average.

Flooding in Germany: 2021
Heavy rainfall events are becoming more intense and frequent.

Heatwaves in India & Pakistan: 2022
Dangerous heat is 30 times more likely.

Typhoon "Hagibis": 2019
Strengthened by climate change, Hagibis caused approximately $4 billion worth of damage to Japan.

Drought in East Africa: 2014
Droughts are becoming more severe and more likely.

Record water temperature in the Central Pacific
Higher temperatures are becoming more likely.

Heatwaves in Australia: 2014
16% higher chance of extreme heat waves.

Large forest fires 2019/20
Fires like this will become more extreme and about 30% more likely.

Current information on extreme weather and its attribution to climate change can be found, for example, here: **worldweather attribution.org**

Sources: CB (2017), Funk et al. (2015), Kam et al. (2015), King et al. (2015), Junghänel et al. (2021), Murakami et al. (2015), Sweet et al. (2013, 2016), Szeto et al. (2016), Shiogama et al. (2013), Zhang et al. (2016), WWA (2020, 2021, 2022)

Future Climate

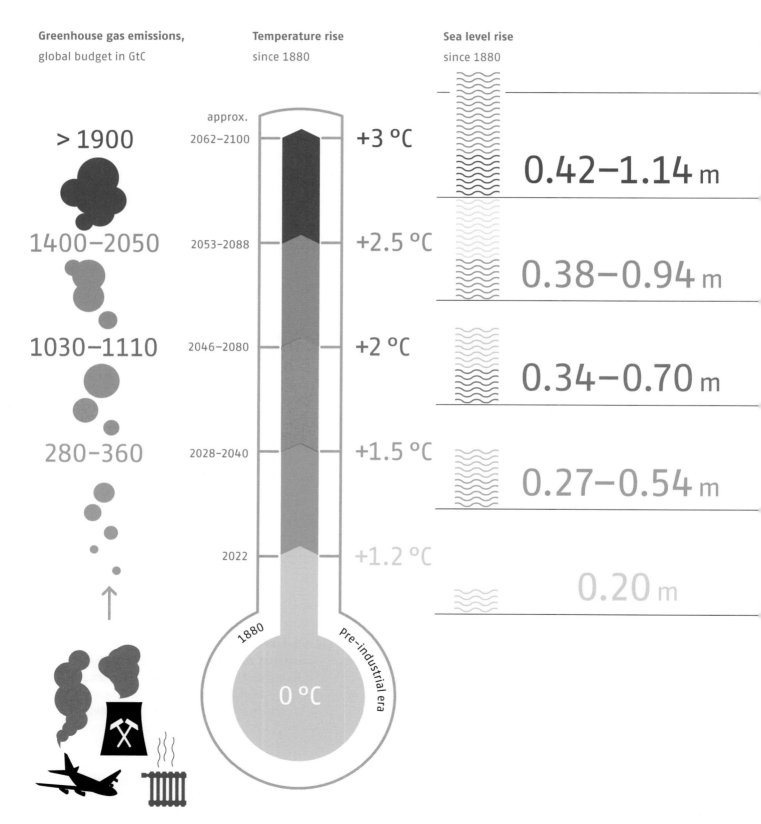

Greenhouse gas emissions,
global budget in GtC

> 1900

1400–2050

1030–1110

280–360

Temperature rise
since 1880

approx.
2062–2100 — +3 °C

2053–2088 — +2.5 °C

2046–2080 — +2 °C

2028–2040 — +1.5 °C

2022 — +1.2 °C

1880

Pre-industrial era

0 °C

Sea level rise
since 1880

0.42–1.14 m

0.38–0.94 m

0.34–0.70 m

0.27–0.54 m

0.20 m

Sources: Eakin et al. (2018), Friedlingstein et al. (2022), IPCC (2021), MCC (2022) , Mora et al. (2017), NASA (2022), WRI (2022)

Coral dieback,
depending on the temperature rise

**Percent of world population
affected by heat waves***
*more than 20 days per year

 100 %

 74% approx. 2062–2100

 67% 2053–2088

 >99%

 61% 2046–2080

 70–90%

 48% 2028–2040

 70%

 30% 2022

**Intergovernmental Panel on Climate Change (IPCC) = worldwide climate council of the United Nations, consisting of over 250 scientists.

If we don't make significant efforts to reduce greenhouse gas emissions, the Earth could warm by almost 3 degrees Celsius by the end of the century. In the worst-case scenario, as projected by the IPCC** climate report, the warming could reach up to 4.4 degrees. Such a rise in temperature would have disastrous consequences for ecosystems, and ultimately, for us humans too.

A 3-degree warmer world would result in widespread flooding, posing a threat to coastlines, river deltas, and islands across the globe. This would cause millions of people to flee their homes. Coral reefs, which are essential to a quarter of all marine life, could die, leading to a collapse of marine ecosystems.

In addition, human food sources would be severely affected by more frequent heat waves, prolonged droughts, and overexploitation. To maintain a livable planet, we must limit the temperature increase to 1.5 degrees Celsius.

Climate Solutions

Food security ①

About 800 million people globally suffer from chronic undernourishment. The situation is becoming even worse due to the ongoing climate crisis. In the world's poorest areas, around 500 million micro-farms produce 80% of the food. To ensure food security, these micro-farms need financial assistance.

Water security ②

In addition to food, it should be ensured that every person in the world has enough clean drinking water. The climate crisis is exacerbating the drinking water shortage in about 20 countries.

Intensify transnational cooperation

International action groups after Climate Summits, for example

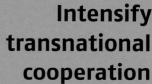

Protection from weather extremes ③

Sea level rise and increasingly frequent extreme weather events such as heavy rain and hurricanes are destroying the homes and livelihoods of countless people, especially in the poorest countries.

Halting deforestation ④

An immediate global halt to deforestation of native forests would have a rapid and enormous effect on the carbon cycle. By absorbing CO_2 and storing carbon, old-growth forests act as climate protectors.

Sources: Bai et al. (2018), Gitz et al. (2016), Scherer & Tänzler (2018)

10 Personal change

.. become a climate hero!
Read more on page 198.

9 Mobility turnaround

An expansion of local transport through trains, electric buses, and cycle paths can help reduce over-all traffic volume, including for the transportation of goods. For example, long freight trains can be used instead of individual trucks.

8 Agricultural turnaround

We should move away from harmful pesticides and monoculture, towards organic farming, permaculture, and healthy soils that store CO_2. Innovative food production methods like aquaponic-towers should be subsidized.

7 Economic turnaround

Top priority should be given to sus-tainability, recycling, green production, and resource conservation instead of expansion and exploitation. The focus should be on "green efficiency" and "zero emissions" targets.

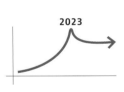

2023

6 Energy and policy shift

The emphasis of subsidies should shift to the expansion of solar energy and efficient storage, while coal and gas-fired power plants ought to pay CO_2 levies instead of receiving subsidies.

5 Smart cities

More than half of the world's population lives in cities, where 75% of global energy emissions are produced. Sustainable green infrastructure conversion, energy production, and home insulation should be top climate goals.

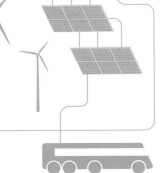

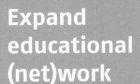

Expand educational (net)work

Online platforms, climate weeks, education, media, TV, books

Four Tools to Minimize Climate Risks

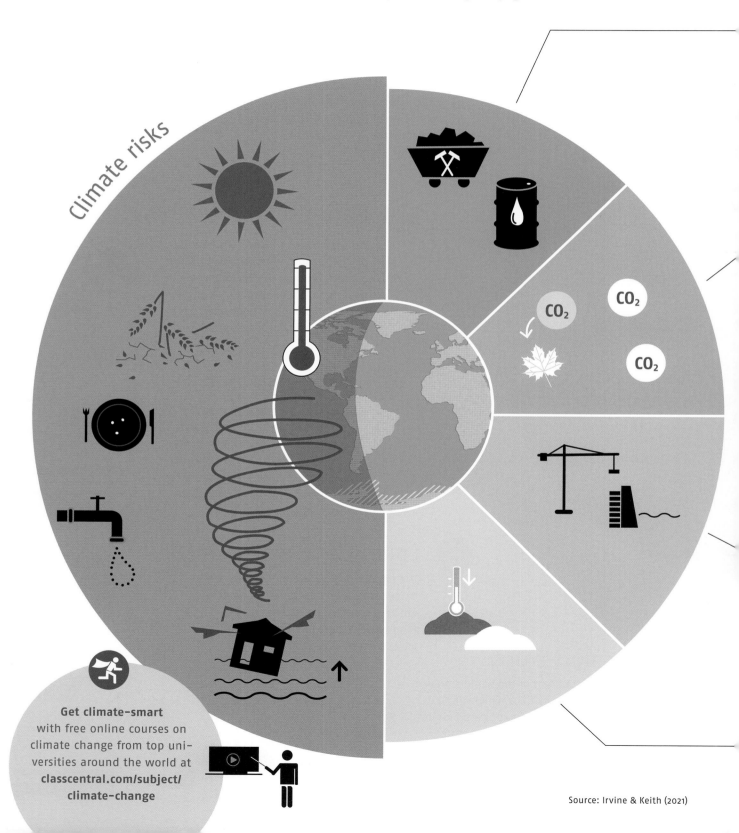

Source: Irvine & Keith (2021)

1 Decarbonization
Stop industrial green-
house gas emissions so
that the greenhouse
effect is not fueled
further. Instead, advance
the transition to renew-
able and carbon-free
energy.
see p. 44–51

2 CO₂ removal
Explore nature-based
and mechanical methods
for extracting CO₂ from
the atmosphere. In
addition, future emis-
sions that are difficult to
eliminate could be offset
by CO₂ removal.
see p. 52/53

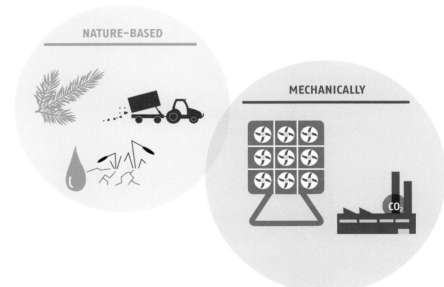

NATURE-BASED

MECHANICALLY

3 Climate adaptation
Prepare societies and
ecosystems to better cope
with the threats of a
changing climate.
see p. 54/55

4 Mitigating climate
impacts with technology
Techniques such as solar
geoengineering could be
used to actively change

the Earth's energy
balance and thus tem-
porarily cool down the
climate.
see p. 56/57

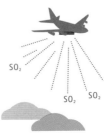

SO₂   SO₂   SO₂

Decarbonization: Global Energy

91%

of the world's population is exposed to air that is harmful to health, according to the World Health Organization (WHO). More green energy also means lower health costs.

89 million

people in Africa and Asia are currently using solar power, with an annual increase of 4%.

58%

of global solar panels are installed in the Asian-Pacific region. Europe and North America combined only reached 34% of global solar power.

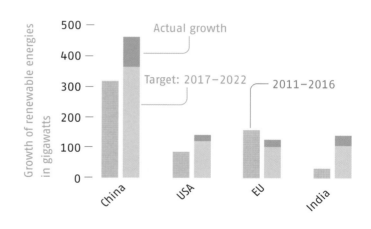

Growth of renewable energies in gigawatts

500
400
300
200
100
0

Actual growth

Target: 2017–2022   2011–2016

China   USA   EU   India

Nuclear

Natural gas

Oil

Coal

Sun

Wind

Water

Geothermal

Biogas

12 million people work in the green energy sector worldwide (2020).

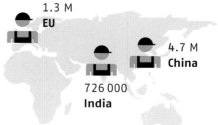

1.3 M
EU

838,000
USA

4.7 M
China

726 000
India

1.2 M
Brazil

Global energy consumption

Electricity, transport & heat (2021)

65% of the energy we use must come from renewable sources by the year 2050, to limit global warming to 2 degrees. However, achieving this target will require the expansion of renewable energy to be carried out 7 times faster than the current rate. As of 2021, the share of renewable energy is about 14% worldwide.

Expansion of renewable energies

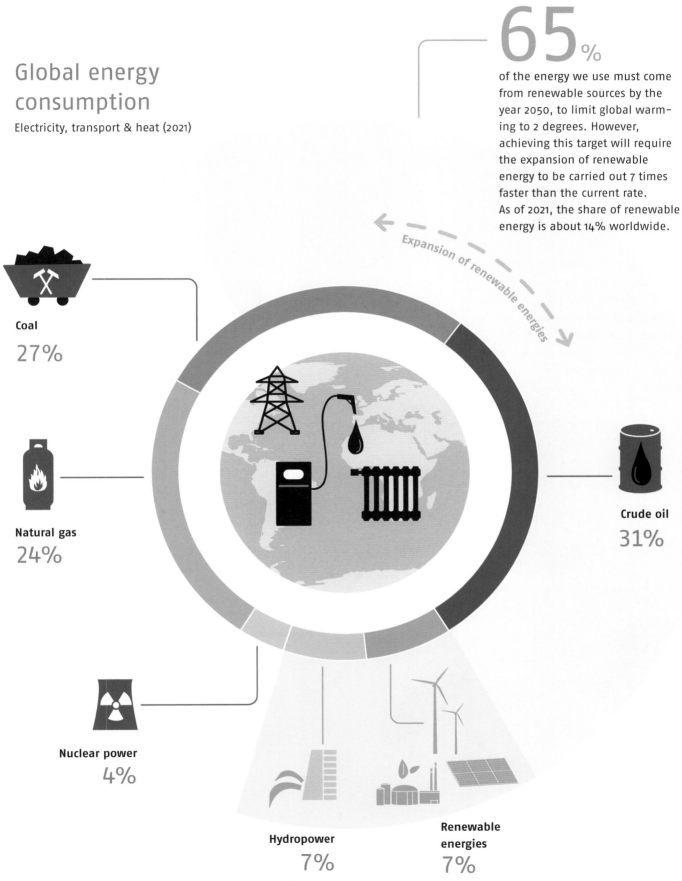

Coal
27%

Natural gas
24%

Crude oil
31%

Nuclear power
4%

Hydropower
7%

Renewable energies
7%

Sources: BP (2022), ISA (2022), IRENA (2021), WBG (2017), WHO (2018)

Instead of Oil and Gas, Use Sun...

Albert Einstein was awarded the Nobel Prize in 1921 for his discovery of the photoelectric effect, which laid the foundation for the development of solar energy.

Solar cells are made up of thin layers of silicon that are readily available from sand. Phosphorus and boron are mixed into the silicon to create the desired properties. The solar panel is typically made up of plastic and glass plates, as well as an aluminium frame. There are also flexible panels made of pliable plastic. It's worth noting that rare earths are not required for the construction of solar cells.

Solar rays consist of many tiny photons.

The alternating current is passed to an electricity meter and then either goes directly into private sockets or is fed into the grid.

The current transformer converts the direct current (DC) into alternating current (AC), that we can use at our sockets.

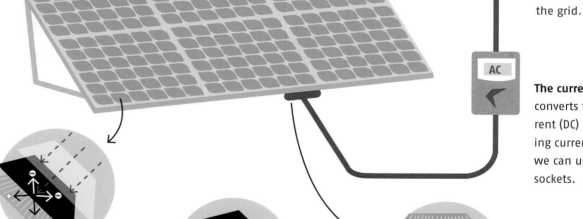

When photons hit the negatively charged top of the silicon layer of the solar cell, the electrons become mobile.

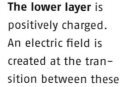

The lower layer is positively charged. An electric field is created at the transition between these two layers.

The electric current flows to the conduction band and through conductors further to the current transformer.

Sources: Quaschning (2021), Rollet (2019)

...and Wind Power

Electricity generation from wind turbines depends on rotor blade length, wind speed, and air density. The longer the blades and stronger the wind, the more energy is produced. Blade length can reach up to 170 meters.

The rotary motion of the rotor blades is multiplied via the gearbox. In this way, kinetic energy becomes mechanical energy.

The generator converts the mechanical energy into electrical energy, like a giant bicycle dynamo.

Nacelle

Depending on the wind direction, the nacelle rotates so that the rotor blades face directly into the wind.

Tower, up to 200 m high

The power lines run through the tower to the ground, where the control technology and the transformer are located.

A transformer at the base of the plant processes the electric current as alternating current for the grid.

The first electricity-generating wind turbine was designed in 1888 by Charles Brush, an American inventor (1849-1929). The turbine was equipped with 144 rotor blades but was only capable of producing 12 kilowatts of power. Later, a Danish scientist named Poul la Cour (1846-1908) discovered that fewer rotor blades moving at higher speeds generated more electricity. In 1904, he started offering courses for wind power engineers, and by 1918, wind turbines were responsible for meeting 3% of Denmark's electricity needs. (Today, the percentage has increased to 52%.) Johannes Juul, one of Poul la Cour's students, is the inventor of the first turbine that generated alternating current, which is considered the forerunner of modern wind turbines. Today's wind turbines generate an average of 3 to 6 megawatts of power.

Sources: Hornung (2020), Fechner & Zwieauer (2021)

Electricity from the Sun vs. Wind

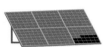

After 1–2
years, solar energy systems have saved more CO_2 than was emitted for their production.

93%
of solar energy output remains after 10 years – just a minimal decrease in efficiency.

10–25%
of typical solar energy output persists when it is cloudy or foggy.

min. lifetime in years

0   10   20   25

After 5–12
months, wind turbines have saved more CO_2 than was emitted for their production.

Wind turbines typically function for 20–25 years, subject to maintenance and environmental factors.

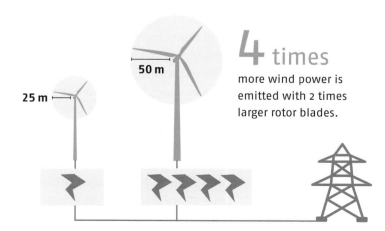

25 m   50 m

4 times
more wind power is emitted with 2 times larger rotor blades.

2%
of Germany's area for wind turbines is sufficient for the green energy transition.

Sources: BWE (2021), Ember (2022), IEA (2022), Kühl (2019)

Solar and wind power share per country (selection, 2021)   ● = 1%

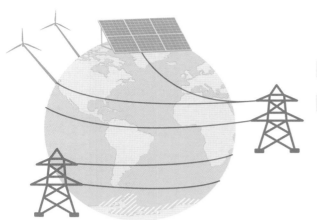

Percentage of solar and wind power in selected countries

52% Denmark

●●●

47% Uruguay

●●●

●●●●●●●●●●●●●●●●●●●●●●●●●●●●●●●●●●● 33% Ireland, Spain

●●●●●●●●●●●●●●●●●●●●●●●●●●●●●●●● 32% Portugal

●●●●●●●●●●●●●●●●●●●●●●●●●●●● 29% Germany

●●●●●●●●●●●●●●●●●●●●●● 22% Australia

●●●●●●●●●●●●●●●●●●●●● 21% Chile

●●●●●●●●●●●●●●●●●● 18% Kenya

●●●●●●●●●●●●● 13% Brazil, USA

●●●●●●●●●●● 11% China

●●●●●●●●●● 10% Japan, Argentina

●●●●●●●●● 9% France

●●●●●●●● 8% India

●●●●●●● 7% Canada

◀ 0.5% Russia, Saudi Arabia

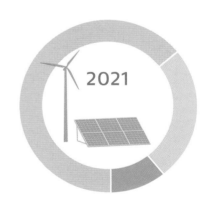

10% of electricity globally came from solar and wind and 38% from all green energies.

Transition to Green Transport

Imported refrigerated produce via cargo planes creates high CO_2 emissions. Stricter import regulations could reduce these emissions.

1 ## Transport emissions
globally, a total of 8.1 Gt CO_2 (2018)

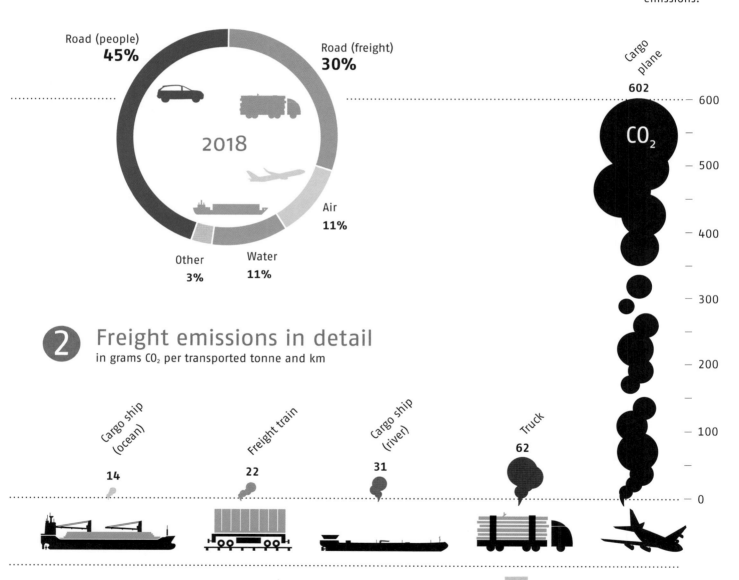

Road (people)
45%

Road (freight)
30%

2018

Air
11%

Other
3%

Water
11%

Cargo plane
602

CO_2

— 600
— 500
— 400
— 300
— 200
— 100
— 0

2 ## Freight emissions in detail
in grams CO_2 per transported tonne and km

Cargo ship (ocean)
14

Freight train
22

Cargo ship (river)
31

Truck
62

+80%
higher emissions in the transport sector since 1990

50%
of global freight emissions are caused by trucks.

By 2050, oil dependence in the transport sector must be drastically reduced to achieve the 2-degree target.

2050

Large cities need to prioritize **cycling infrastructure,** following the example of the Netherlands and Denmark.

*incl. energy related emissions

The transport sector is currently the fastest-growing source of CO_2 emissions globally, and it has the largest potential for savings. Thus, it is one of the top priorities for climate protection. About 24%* of the world's greenhouse gases are emitted in the transport sector, and this figure is rising by 2.5% annually. China has experienced the sharpest increase since 1990.

In the freight sector, both the transportation of goods and emissions are expected to increase up to four times by 2050 compared to 2010. Transport of both passengers and freight by road has a significant impact on climate emissions. However, emissions can be reduced by providing alternative public electric transportation and more e-freight transport, such as by train or e-truck. The consumption behavior of every individual also affects freight transport.

Public transport must be expanded and should be offered free of charge to reduce the number of cars in cities.

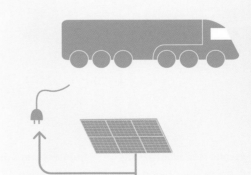

E-trucks are already in use and are suitable for urban and regional transport. At the moment, long distances remain a challenge.

Currently, China leads the world in **electric car usage,** followed by Europe and the USA. However, the production of electric car batteries is highly energy- and resource-intensive.

>10 million

E-vehicles are on the roads worldwide (as of 2020) and this number is increasing.

Sources: IPCC (2022), IEA (2022), Paoli et al. (2021), Tiseo (2021), Plötz & Link (2021)

2 Binding CO$_2$: Naturally vs. Artificially

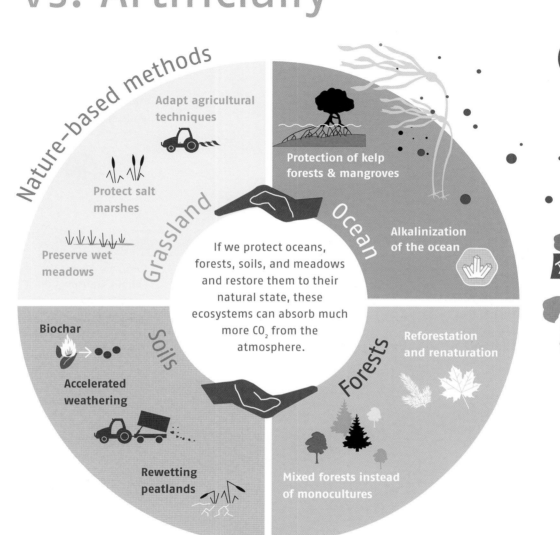

Nature-based methods

Adapt agricultural techniques

Protect salt marshes

Preserve wet meadows

Biochar

Accelerated weathering

Rewetting peatlands

Grassland

Soils

If we protect oceans, forests, soils, and meadows and restore them to their natural state, these ecosystems can absorb much more CO$_2$ from the atmosphere.

Protection of kelp forests & mangroves

Alkalinization of the ocean

Reforestation and renaturation

Mixed forests instead of monocultures

Ocean

Forests

Biochar

When biomass is converted to biochar, approximately one-third of the CO$_2$ absorbed by the plant is retained in the coal. This carbon can be stored in the soil for an extended period of time when used as a fertilizer. Additionally, the biochar acts as a sponge in the soil, retaining water and nutrients.

Accelerated weathering & rewetting of peat

Fertilizing soil with rocks like basalt can help increase its CO$_2$ fixation. When these rocks come into contact with water, they absorb CO$_2$ from the air. Additionally, drained peatlands can effectively absorb more CO$_2$ than they emit after being re-wetted.

Reforestation and renaturation

Forests store around a quarter of our annual CO$_2$ emissions. Naturally mixed forests have a much higher carbon storage capacity than tree plantations. Reforestation and restoration of degraded and fragmented forests in the tropics are particularly effective.

Ocean-based solutions

Oceans store approximately 25% of carbon dioxide emissions. To increase CO$_2$ absorption in the ocean, minerals could be added to the seawater to increase its alkalinity, up- or downwelling could be artificially created, as well as macroalgae farms and seagrass established.

Sources: Carbfix (2022), EnergyNow Media (2021), Geoengineering Monitor (2021), GEOMAR (2021), IPCC (2021), Mengis & Kalhori (2021) UC Davis (2019)

Artificial CO_2 extraction from the air

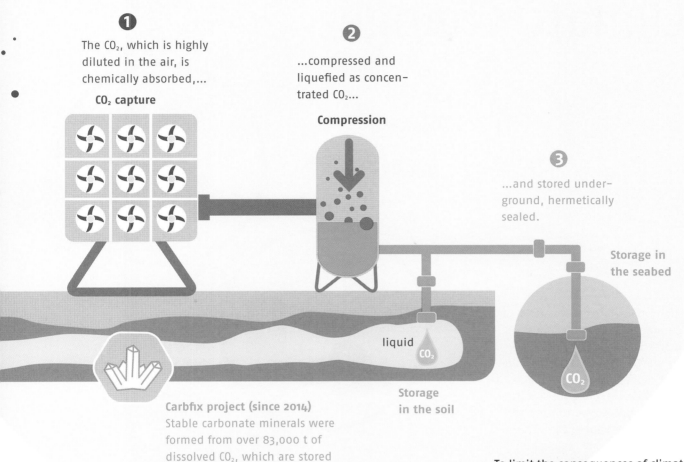

1 The CO_2, which is highly diluted in the air, is chemically absorbed,...

CO_2 capture

2 ...compressed and liquefied as concentrated CO_2...

Compression

3 ...and stored underground, hermetically sealed.

Storage in the seabed

liquid CO_2

Storage in the soil

CO_2

Carbfix project (since 2014)
Stable carbonate minerals were formed from over 83,000 t of dissolved CO_2, which are stored underground.

CO_2 removal in a biomass power plant?

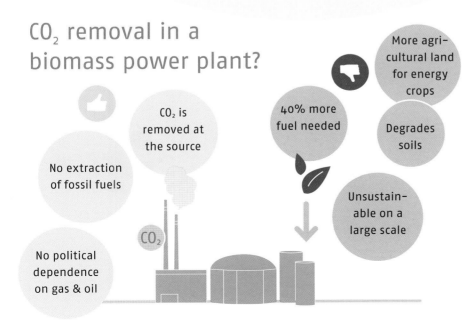

CO_2 is removed at the source

No extraction of fossil fuels

No political dependence on gas & oil

CO_2

40% more fuel needed

More agricultural land for energy crops

Degrades soils

Unsustainable on a large scale

To limit the consequences of climate change to a maximum of 2 degrees Celsius, it is imperative that we cease the use of fossil fuels as soon as possible. The longer we delay our transition to alternative energy sources, the more CO_2 we will have to extract from the air.

Currently, CO_2 capture plants under construction have the capacity to remove around 3 million tonnes (Mt) of CO_2 per year from the atmosphere. In addition, there are plans to build CO_2 capture plants with a capacity of 108 Mt as of September 2021, but to achive the 2-degree goal, we would need an annual capacity of 5600 Mt by 2050.

Various pilot and research projects are underway to investigate how CO_2 can be captured directly at biomass power plants or other sources. However, biomass plants remain problematic because of their environmental impact and lack of large-scale sustainability.

③ Adaptation to Climate Change

Health problems
Without adaptation:
maximum high risk
With high adaptation:
high risk

Polar regions

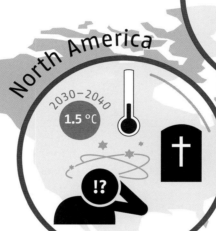

2080–2100
4 °C

North America

2030–2040
1.5 °C

Swift action is imperative if we are to keep global warming to acceptable levels. To meet the 1.5–degree target, global CO_2 emissions must be halved by 2030. Additionally, all countries must invest more in adaptation strategies to reduce or mitigate existential risks for people. This includes building and strengthening dams, protecting dunes and coastal landscapes, safeguarding drinking water supplies, protecting soils from degradation, and using drought–resistant species to ensure food security.

It is crucial to protect the remaining wilderness that is still intact if we want to safeguard essential livelihoods for everyone in the future. Biodiversity, coral reefs, dune landscapes, mangroves, glaciers, and forests all have a significant impact on food and drinking water security, as well as climate adaptation, but only when they remain undisturbed.

Increase in deaths from heat waves
Without adaptation:
high risk
With high adaptation:
low risk

South America

2030–2040
1.5 °C

Help coastal dwellers!
Find out where urgent help is needed on the interactive world map of future sea level rise: **https://sealevel.nasa.gov/data_tools/17**

Decrease in crop yields and food security
Without adaptation:
maximum high risk
With high adaptation:
low risk

Possible temperature **increases** of 1.5 to 4 degrees over different time periods

Europe

Growing damage from river floods and coastal flooding
Without adaptation: medium to high risk
With high adaptation: low risk

Growing damage from extreme heat and forest fires
Without adaptation: maximum high risk
With high adaptation: medium to high risk

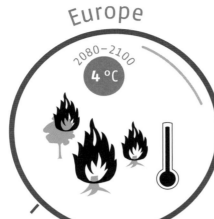

Europe

Asia

Increasing damage to infrastructure and dwellings due to flooding
Without adaptation: high risk
With high adaptation: medium risk

Africa

Decreased crop yields, drinking water, and food security
Without adaptation: high risk
With high adaptation: low risk

Australia + Oceania

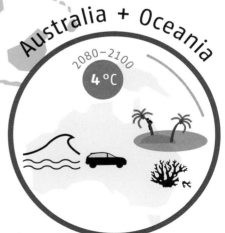

Damage to coastal infrastructure, ecosystems, and islands
Without adaptation: high risk (maximum for islands)
With high adaptation: medium risk (high for islands)

Sources: IPCC (2014), UNFCCC (2017)

4 Mitigating Climate Impacts...

Four ideas for reducing solar radiation

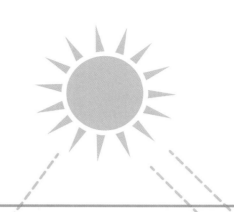

Stratospheric aerosol injection

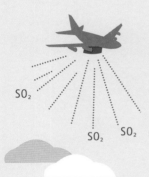

SO₂

SO₂   SO₂

During large volcanic eruptions, immense amounts of aerosols* are ejected into the atmosphere where they reflect sunlight, leading to a temporary cooling of the climate. This could be imitated by spraying sulphur dioxide (SO_2) from aircraft into the stratosphere.

However, this technique is only effective during sunny weather, and it has the strongest impact in tropical regions, where it could lead to significant changes in the water cycle. Additionally, if the aerosol injection is stopped, it could intensify global warming within a few years.

Brightening of the clouds

Cloud cover could be lightened up and enlarged with sprayed seawater so that more sunlight is reflected into space (albedo effect). This could reduce the temperature of the atmosphere and the oceans.

However, there have been no successful experiments on a large scale so far, and negative side effects such as altered rain patterns need to be further investigated. This method would have to be applied continuously, so the chances of global implementation are poor.

Thinning of the cirrus clouds

Cirrus clouds are high, very cold clouds of ice crystals. These clouds act like greenhouse gases: heat radiation that is radiated back from the Earth towards space is blocked by the cirrus clouds.

With the help of drones, fine dust particles could be sprayed onto these clouds to thin them out or dissolve them completely, leading to more heat radiation being radiated back into space.

However, the opposite can also happen: If there are too many particles, the clouds would become *thicker*!

*Floating particles, the smallest particles are a few nanometers in size.

Sources: Burns et al. (2019), Hüttmann (2019, 2020), Irvine & Keith (2020), Irvine (2022)

...with Solar Geoengineering?

Reflectors in space

Another idea is to use reflectors in space to prevent a small part of the sun's radiation from hitting the Earth by reflecting it.

The implementation of this idea would be very expensive because the large reflectors would have to be launched into space with rockets. It would take at least 25 years to develop such a technique. Therefore, it will probably remain a theory.

Pro

Some scientists and politicians argue that geoengineering should be used in parallel with emissions reduction to temporarily mitigate the negative consequences of climate change: to reduce the temperature rise in the short term, to protect coral reefs selectively and, for example, to prevent Arctic sea ice from melting. They argue that the desired cooling effect could be achieved relatively quickly compared to emission reductions based on behavioral changes.

Con

Geoengineering opponents, however, warn of the potential negative consequences for the environment that have not yet been researched, difficulties in regulating use worldwide, and new climate impacts that could arise regionally and which are not visible in risk calculations with global climate models.

Efficiency and affordability over a longer period are not guaranteed, and social injustice in the climate crisis would be exacerbated, especially in the global South.

What is the albedo effect?

The surface of the Earth can be categorized according to its reflectivity (albedo). Up to 90% of solar radiation is reflected from white surfaces such as fresh snow, while the open sea reflects the lowest amount, at 7%. Ideas for solar geoengineering are based on the albedo effect.

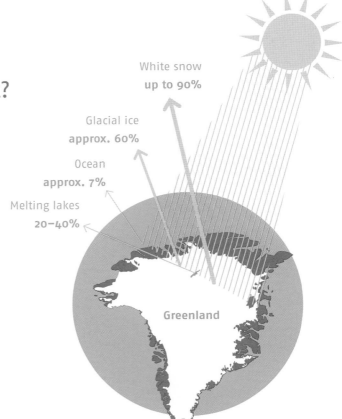

White snow
up to 90%

Glacial ice
approx. 60%

Ocean
approx. 7%

Melting lakes
20–40%

Greenland

THE HYDROSPHERE

Water, Ice & Snow

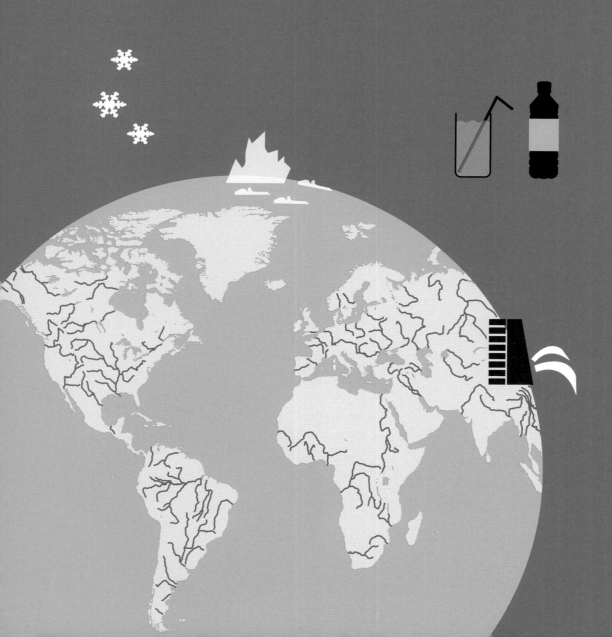

Hy | dro | sphere
[Ancient Greek *hýdor* = water
& *sphaĩra* = sphere (Earth)].
All the water on Earth together forms the
hydrosphere, regardless of its state:
liquid in water bodies, and
oceans, gaseous in the air and in clouds,
or frozen in sea ice and snow.

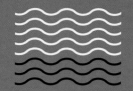

Saltwater

The Seas of the Earth

ARCTIC OCEAN

ATLANTIC OCEAN

PACIFIC OCEAN

SOUTHERN OCEAN

5 basins

form the global ocean: the Atlantic, Pacific, Indian, Arctic, and Southern Oceans. In addition, there are larger and smaller seas that are enclosed or bordered by land masses, such as the European Mediterranean or the Black Sea.

3.5%

of the weight of seawater consists of dissolved salt. The salt content is lower at the equator and the poles than in the mid-latitudes.

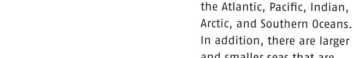

Approx.

91%

of the estimated 2 million animal and plant species in the oceans are unexplored and unknown.

Only **9%** of all marine species are identified.

Sources: Gartner & Armstrong (2012), GEOMAR (2021), Mora et al. (2011), NOAA (2021)

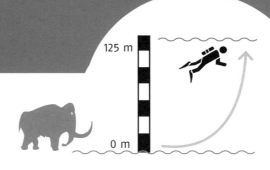

INDIAN
OCEAN

Approx.
70%
—— of the Earth's
surface is
covered by
oceans.

125 m

0 m

125 m
higher than the
sea level during the
last ice age.

4 x more
oxygen–depleted dead
zones. Global warming
has reduced the oxygen
content of the oceans by
40% since 1970.

SCIENCE

Around
11,000 m
is the deepest part
of the ocean, the
Mariana Trench in
the Pacific.

Only **20**%
of the seabed has been explored
and mapped in detail with
echo sounders.

**Get involved with
marine biology!**
Find over 150 free online courses
on ocean topics, made by the
world's leading universities:
classcentral.com/search?q=ocean

Humanity...

We profit from
the oceans
and use them as a:

Food provider
Fish, algae, and bivalves
constitute the staples
of many people's diets.

Poverty reducer
In many developing countries,
fish provides the only
affordable source of protein.

**Energy and
resource supplier**
from petroleum
to offshore wind farms.

Employer
Up to 12% of all employed
people worldwide are
dependent on the fishing
industry.

Transportation route
Billions of goods are
transported across
the oceans annually.

Medication provider
Several medications are
produced with substances
extracted from the ocean.

How
humanity
threatens
the oceans:

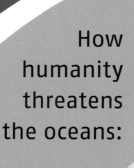

Climate change

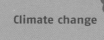

Pollution

Industrialization

Overfishing

Oasis of rest and recreation
Beaches and coastal regions
are popular recreational and
vacation destinations.

063

...and the Sea

The oceans function as:

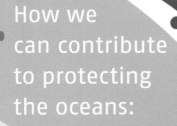

How we can contribute to protecting the oceans:

Climate regulators
They control the weather in their constant exchange with the atmosphere.

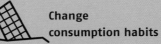

Change consumption habits

Habitats
They facilitate a balanced biosphere through complex food chains.

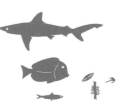

More recycling, less plastic

Climate protectors
They provide a "buffer action" by absorbing CO_2 and heat from the atmosphere.

Reduce CO_2 emissions

Oxygen providers
The photosynthetic plankton and other plants in the ocean produce 50% of the oxygen we breathe.

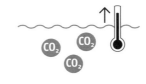

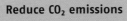

Eat less fish

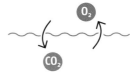

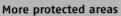

More protected areas

Stabilizers
They promote biodiversity by providing stable ecosystems.

Nurseries
Coral reefs provide a safe space for reproduction and biodiversity.

Source: HBS (2017)

Oceans and Climate Change

1 The ocean is warming

Sea surface temperatures are rising due to climate change. The deeper layers of the seas are also warming, but much more slowly. Compared to 1980, the oceans are about 0.6 degrees Celsius warmer today (2020), while the tropics are warming faster than other regions.

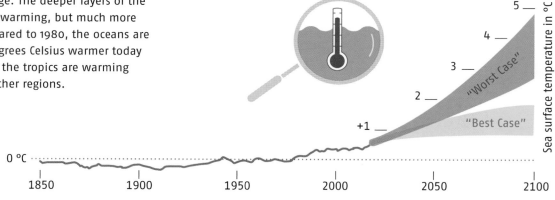

2 Marine biodiversity is declining

Corals engage in a symbiosis with certain algae, the zooxanthellae, which are indispensable to life. Corals are fed by the algae and gain their color through them.

Starting at a 1°C rise in temperature, the algae go into a state of shock and produce toxins instead of sugars. The coral then rejects its partners and loses its color.

As a result, the corals starve. After their death, a dangerous process of algae and sponge coverage begins, making the return of zooxanthellae virtually impossible.

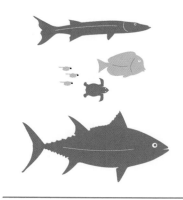

25%

of the ocean's marine life is directly dependent on coral reefs

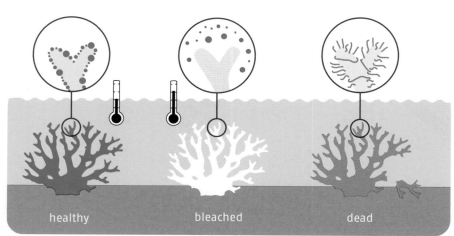

healthy   bleached   dead

Sources: ARC (2022), IPCC (2021), NOAA (2019)

3 Acidity levels are rising

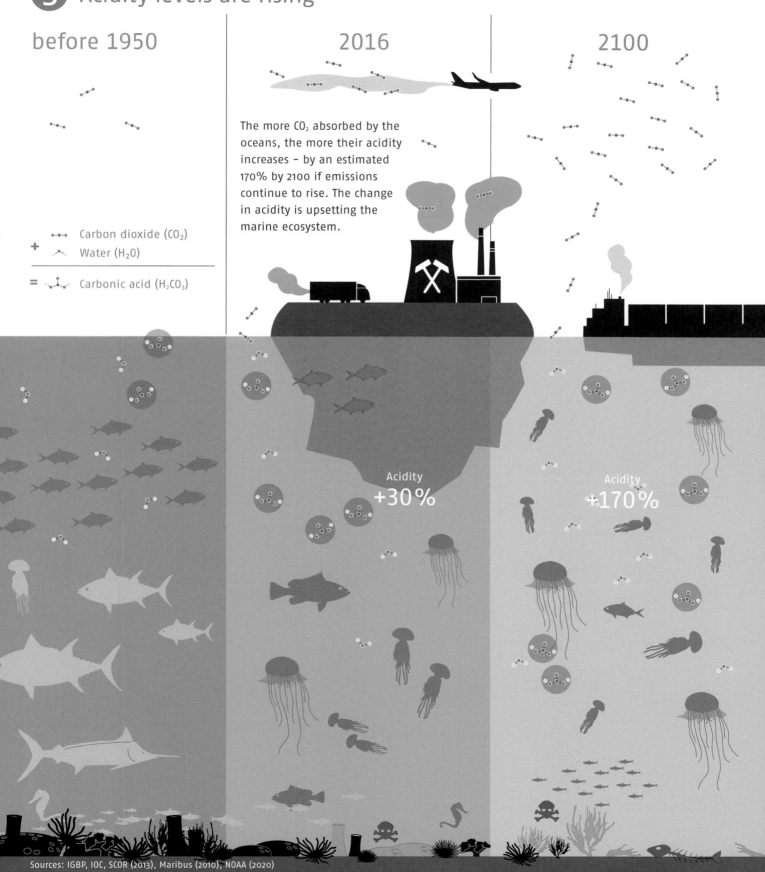

before 1950

2016

2100

+ ·•·•· Carbon dioxide (CO₂)
   ·∧· Water (H₂O)

= ·∪· Carbonic acid (H₂CO₃)

The more CO₂ absorbed by the oceans, the more their acidity increases – by an estimated 170% by 2100 if emissions continue to rise. The change in acidity is upsetting the marine ecosystem.

Acidity
+30%

Acidity
+170%

Sources: IGBP, IOC, SCOR (2013), Maribus (2010), NOAA (2020)

Garbage in the Sea

Plastic is now everywhere – even in places we never go. It is accumulating in the world's oceans in 5 giant gyres. The visible garbage floating on the surface can be compared to the "tip of the iceberg," as the slurry of small and large plastic parts in a garbage gyre extends about thirty meters into the depths. The sun's rays, the salinity of the water, and the constant movement cause the plastic to decompose at different rates: it can take between 1 and 600 years for plastic bags or fishing lines to decompose into pieces the size of grains of sand. Much of the waste eventually sinks to the bottom of the ocean and is eventually covered by sediment. The highest density of plastic on the seafloor has been measured in Indonesia, with about 690,000 particles per square kilometer.

For every kilogram of zooplankton in the North Atlantic Gyre, there are 6 kg of plastic floating around.

5 large garbage gyres are formed by the ocean currents.

Sources: Eriksen (2014), GP (2007), ICC (2010), Moore (2001), Maribus (2010), UNEP (2005)

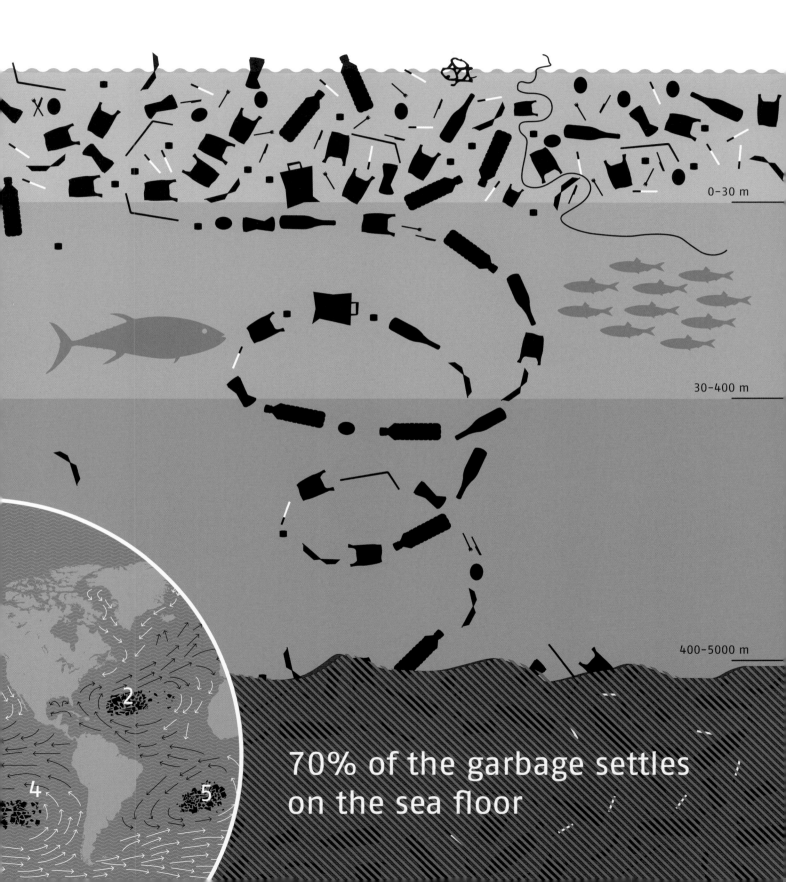

0–30 m

30–400 m

400–5000 m

70% of the garbage settles on the sea floor

Freshwater

Rivers and Dams

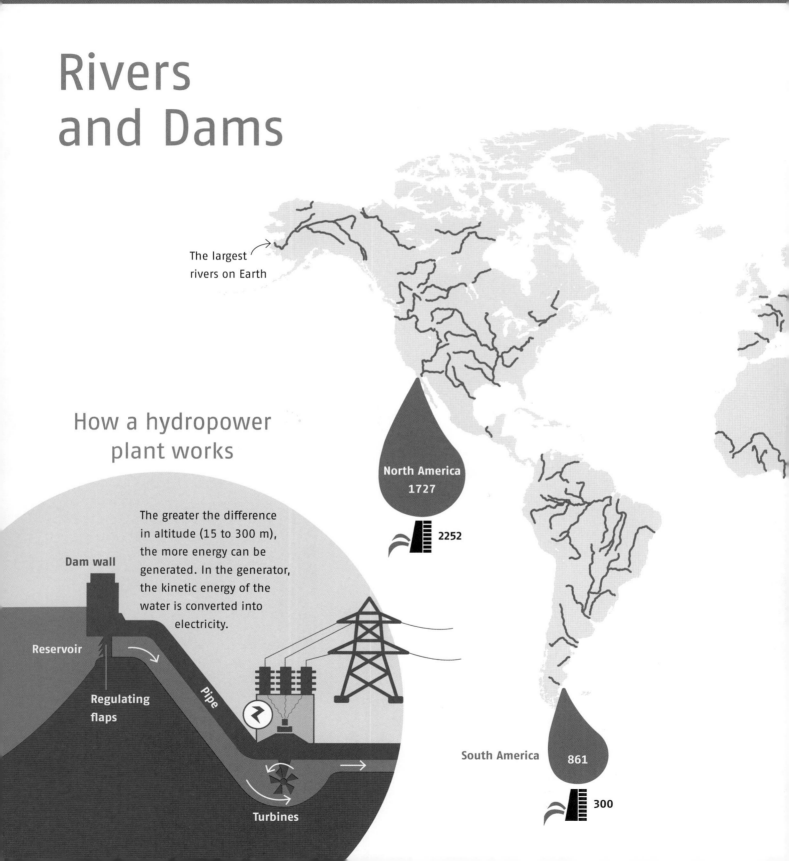

The largest rivers on Earth

How a hydropower plant works

The greater the difference in altitude (15 to 300 m), the more energy can be generated. In the generator, the kinetic energy of the water is converted into electricity.

Dam wall

Reservoir

Regulating flaps

Pipe

Turbines

North America
1727

2252

South America   861

300

Dams per continent

 Reservoir volume in billion m³    Number of dams

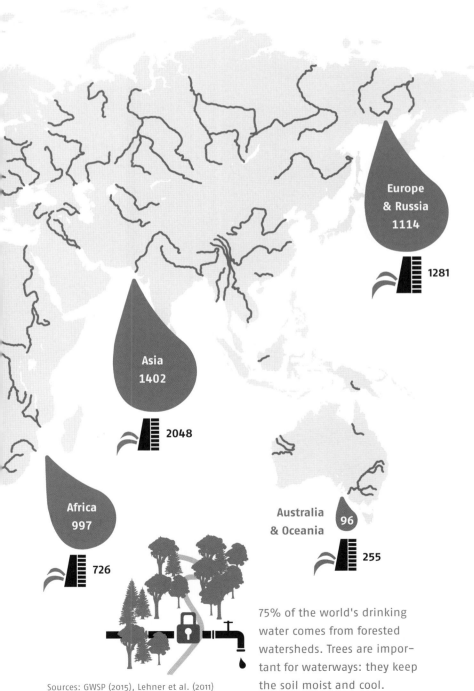

Europe & Russia
1114
1281

Asia
1402
2048

Africa
997
726

Australia & Oceania
96
255

75% of the world's drinking water comes from forested watersheds. Trees are important for waterways: they keep the soil moist and cool.

Sources: GWSP (2015), Lehner et al. (2011)

Rivers are like lifelines, essential for food, drinking water, and culture. Rivers are also used to transport goods, generate electricity, cool industrial plants, and irrigate fields.

But rivers also carry waste, chemicals from the textile industry, pesticides from the fields, mercury from gold mining or oil, and metals from mines.

About half of the world's rivers have at least one dam, whether for flood control, drinking water supply, or power generation. Reservoirs are also hugely important for agriculture: 30–40 percent of all irrigated land is supplied with water from reservoirs. The construction of dams destroys habitats. In Southeast Asia in particular, the rainforest is being cleared for dam projects, with dramatic consequences for many endangered species.

Climate change is altering rivers on every continent. Heatwaves cause oxygen depletion and fish kills, and melting glaciers lead to low water levels. In the short term, the melting of glaciers will increase the output of dams, but in the long term, energy production will decline.

To ensure food production and mitigate the effects of global warming, it is essential to preserve forested river ecosystems through restoration projects and protected areas.

Hidden Freshwater

Deep in the ground: aquifers

Only 2.5% of the world's water is fresh-water. One-third of this is groundwater, and almost two-thirds is stored in ice sheets.

Groundwater feeds springs and supplies rivers, lakes, and forests. It is vital to us in many ways: as drinking water, as the basis for growing food, for manufacturing goods and, in some regions, for heating energy from thermal springs.

Water is the most used resource on the planet. Because of its importance, the quality of groundwater should be constantly monitored. But even in the United States, groundwater is not of sufficiently high quality in many places and has to be treated at great expense (see p. 95).

Extending the legal ban on persistent, non-biodegradable pollutants*, which are still used in pesticides and outdoor clothing, for example, and stricter regulation of over-fertilization in agriculture would be important steps towards ensuring clean groundwater for future generations.

It can take several hundred years for water to percolate from the surface to deep groundwater. So we need to start protecting tomorrow's water today!

*persistent organic pollutants (POPs)

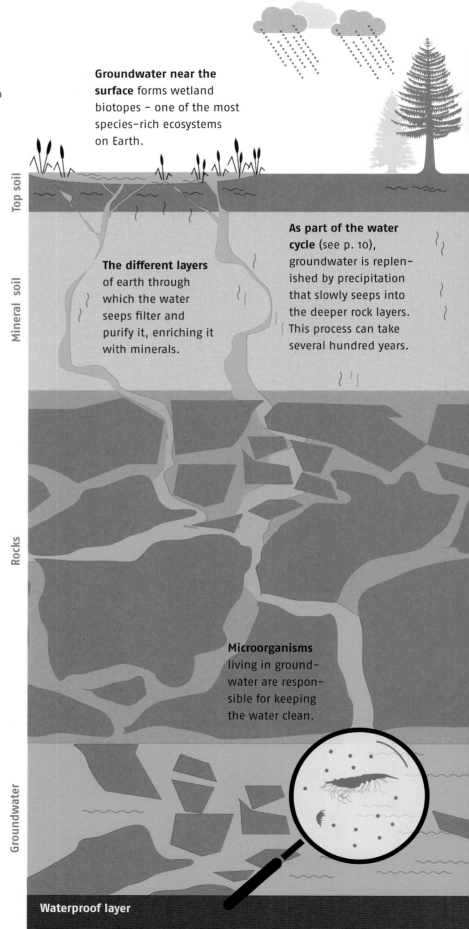

Groundwater near the surface forms wetland biotopes – one of the most species-rich ecosystems on Earth.

The different layers of earth through which the water seeps filter and purify it, enriching it with minerals.

As part of the water cycle (see p. 10), groundwater is replenished by precipitation that slowly seeps into the deeper rock layers. This process can take several hundred years.

Microorganisms living in groundwater are responsible for keeping the water clean.

Top soil

Mineral soil

Rocks

Groundwater

Waterproof layer

Uses of groundwater

Drinking water	Industry	Agriculture	Thermal springs and geothermal energy

About half of the world's drinking water comes from groundwater. In the US, this number is about 26%.
It is also bottled as mineral water.

In factories and power plants, water is used as a coolant or in product manufacturing, such as paper.

Groundwater is used in agriculture to irrigate crops and livestock.

Hot and warm healing springs are used for health reasons and as geothermal energy for heating.

Sea level today

Deep beneath the ocean: freshwater reservoirs

During the last Ice Age, the sea level was about 125 meters lower than it is today. As the Earth warmed and ice sheets 1000 meters high melted, sea levels rose.

A huge freshwater reservoir has been discovered under the sea on the shallow continental shelf off the east coast of the USA, a relic of the Ice Age. The reservoir might continue to be fed by freshwater sources on land while it is also slowly seeping into the sea through sediments.

Sea level during the last Ice Age

Sources: Gustafson et al. (2019), Li et al. (2021), USGS (2018)

Wetlands for the Climate

Wetlands are unique habitats because they form a transition zone between terrestrial and aquatic ecosystems. They are also some of the Earth's most productive and biodiverse landscapes. Highly adapted plants, amphibians, fish, algae, and many bird species form a complex ecosystem that has been called a "biological super-system." Many species are threatened by the fast disappearance of their waterlogged habitat.

Wetlands include swamps, bogs, peat bogs, mangroves, and salt marshes. They have many things in common: their soils contain a lot of water, have a low oxygen content and, from a global perspective, absorb huge amounts of greenhouse gases.

Protect the moor!
Many inspiring initiatives can be found under the hashtag **#WorldWetlandsDay**

Peat swamp forests

A peat swamp forest is a tropical lowland rainforest that grows on up to 30-meter thick wet bog layers.

A healthy peat bog ecosystem has a high water table, the level at which underground soil is saturated with water.

In Southeast Asia, peatland forests are being felled for timber production, for example, or destroyed by illegal slash-and-burn.

If the forest disappears, the soil starts to drain, and CO_2 and methane are released.

Instead of forests, oil palms are planted to meet the increasing global demand for palm oil.

After this point, heat and a lack of soil moisture cause fires and further CO_2 is released.

In its final stage, the peat bog system is destroyed, and the groundwater level is at its lowest limit.

Gradually, the water table sinks, and the moor dries out, and releases the stored CO_2.

Marshes & salt marshes

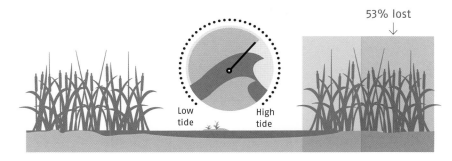

53% lost

Low tide — High tide

Marsh soil is mineral-rich, moist, and densely overgrown with a variety of grasses and reeds.

Periodic flooding, such as by tides, ensures a high nutrient content.

In the US, more than half of the nation's wetlands were lost between 1780 and 1980. Recent conservation efforts led to a net gain of 12,000 ha annually (1998-2004).

Coastal salt marshes are frequented by freshwater- and saltwater-loving species.

Condition of the peatlands

Of the
4 million km²
peatlands
worldwide...

1560 million t CO_{2e}* —

Annual emissions from drained peatlands worldwide and the main polluters (as of 2020)

Indonesia
650

EU
230

Russia
200

Rest of the world
480

...about 500,000 km² are degraded and drained.

Approx.
5000 km²
of intact peatlands are drained, degraded, or destroyed every year. This is an area twice the size of Luxembourg. As a result, former peatlands are emitting greenhouse gases.

*CO_2 equivalents: greenhouse gases converted to the equivalent amount of CO_2 emissions.

Sources: Dahl (2006), Hooijer et al. (2010), Page et al. (2011), Joosten (2022)

 # World of Ice

The Earth's Cryosphere

The cryosphere, from the ancient Greek *krýos* = ice cold and *sphaira* = sphere (Earth), is the totality of all ice and snow surfaces on a planet. On Earth, it covers more than a hundred countries in the form of ice sheets, snow, glaciers, sea ice and ice shelves, permafrost on land and the seabed, and ice on lakes and rivers.

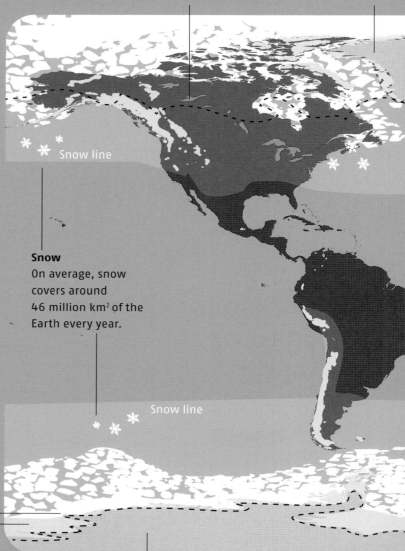

Frozen freshwater
The upper layer of lakes and rivers is often frozen in winter.

Greenland Ice Sheet
Greenland's ice masses are up to 3200 m thick.

Snow line

Permafrost
In the Northern Hemisphere, the ground freezes all year round in parts of Alaska, northern Canada and Siberia, as well as in high mountain regions such as Tibet. In the Southern Hemisphere, small areas of permafrost are found in New Zealand and South America, while the largest area is in Antarctica.

Snow
On average, snow covers around 46 million km² of the Earth every year.

Snow line

Ice shelf
Floating areas of permanent ice, several hundred meters in height, which are in contact with the ice sheets on land.

With 473,000 km², the Ross Ice Shelf in Antarctica is the largest of its kind. Icebergs form when the ice shelf calves at the edges.

Ice sheets
Perennial ice sheets on land that are at least 50,000 km² in size and have a curved shape. The ice moves under its own weight from the centre of the ice dome to the edges.

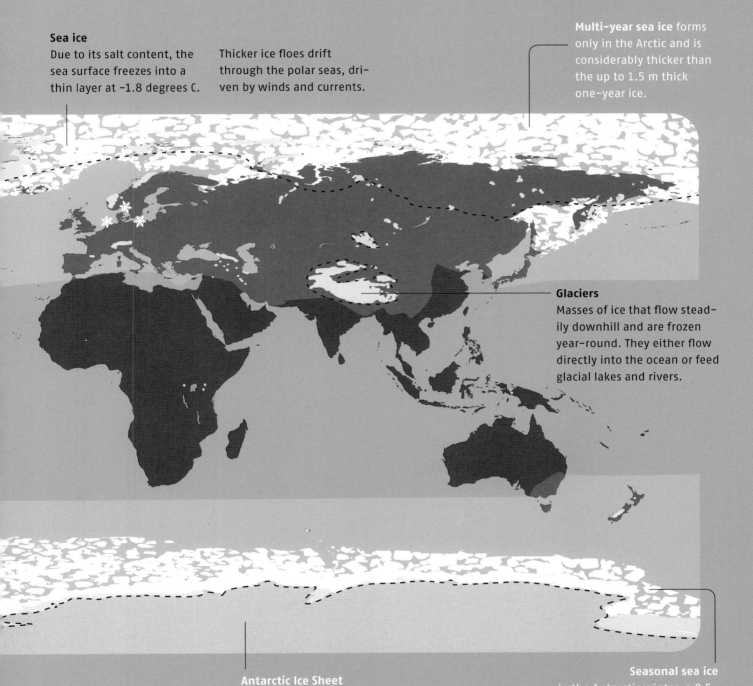

Sea ice
Due to its salt content, the sea surface freezes into a thin layer at −1.8 degrees C.

Thicker ice floes drift through the polar seas, driven by winds and currents.

Multi-year sea ice forms only in the Arctic and is considerably thicker than the up to 1.5 m thick one-year ice.

Glaciers
Masses of ice that flow steadily downhill and are frozen year-round. They either flow directly into the ocean or feed glacial lakes and rivers.

Antarctic Ice Sheet
The ice masses of the Antarctic are up to 4897 meters thick.

*In the winter of 2023, sea ice extent was at a record low, 1.5 million km² less than the average.

Seasonal sea ice
In the Antarctic winter, a 0.5–1 m thin sea ice cover forms*. In summer, only about 20% of the sea ice remains. The reason for this, apart from higher temperatures, is the strong offshore winds.

Sources: Armstrong et al. (2019), NOAA (2021), SSEC (2017), Wadhams (2016), WMO (2021), WOR (2019)

What is Ice?

Ice takes many forms: At sub-zero temperatures, ice flakes form in the air, ice crystals grow in the soil, and ice needles float on the water.

From snowflakes...

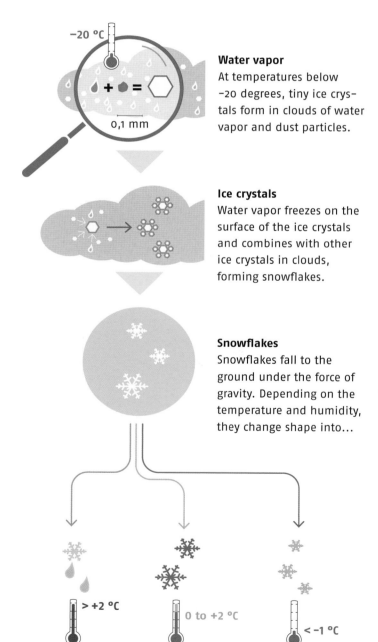

Water vapor
At temperatures below −20 degrees, tiny ice crystals form in clouds of water vapor and dust particles.

Ice crystals
Water vapor freezes on the surface of the ice crystals and combines with other ice crystals in clouds, forming snowflakes.

Snowflakes
Snowflakes fall to the ground under the force of gravity. Depending on the temperature and humidity, they change shape into...

> +2 °C	0 to +2 °C	< −1 °C
Sleet or rain	Large, wet snowflakes	Fine, dry powder snow

...to glacial ice

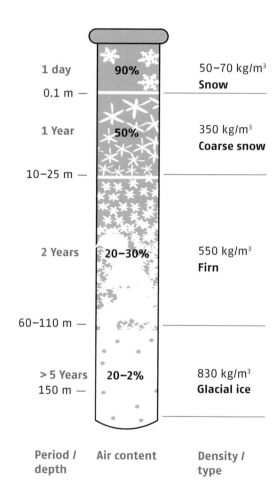

Period / depth	Air content	Density / type
1 day / 0.1 m	90%	50–70 kg/m³ — Snow
1 Year	50%	350 kg/m³ — Coarse snow
10–25 m		
2 Years	20–30%	550 kg/m³ — Firn
60–110 m		
> 5 Years / 150 m	20–2%	830 kg/m³ — Glacial ice

Compressed by its weight and by repeated melting and freezing, after about five years, light new snow becomes heavy glacier ice. A few air bubbles remain in the glacier ice, but pressure reduces their number until only about 2% remain.

Researchers can read the climate history of the last 800,000 years in these bubbles.

Ice is always on the move

Accumulation
Glaciers and ice sheets grow each year as a result of snowfall. Snow and ice avalanches and the refreezing of meltwater also contribute.

Ablation
Glaciers and ice sheets lose mass each year through melting, runoff, sublimation*, calving, and ice breaking off from the ice shelf or glacier front.

*solid ice turns into water vapor

Ice & atmosphere
During ice formation and melting, air temperature and humidity determine whether clouds, snow, and later, ice form. Solar radiation, CO_2 levels, and the transfer of heat between ice, ocean, and atmosphere all contribute to melting.

Glacier movement
The ice in glaciers and ice sheets is constantly flowing downwards under the force of gravity due to its enormous weight. As this occurs, the mass of ice inside the glacier deforms (a phenomenon known as creep).

Ice & oceans
The areas covered by sea ice vary greatly with the seasons. Sea ice moves with ocean currents and winds, changing the seawater's salinity and acting as an insulator between the ocean and the atmosphere.

Ice & bedrock
On the underside of an ice sheet or glacier, the ice masses slide on a thin film of water. This water forms due to the enormous pressure and higher temperatures caused by geothermal heat and the roughness of the bedrock (frictional heat).

Sources: Garthwaite (2019), Kornhuber et al. (2019), NOAA (2019), Pruppacher & Klett (1978), WOR (2019)

Ice from the Sea

Saltwater

The freezing of seawater creates what is known as sea ice.

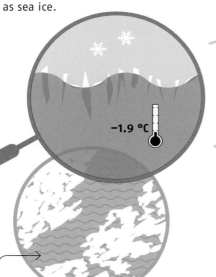

−1.9 °C

In "polynyas", ice-free areas in the polar oceans and on the coasts, cold wind cools the water surface. This is how ice crystals are formed – first in needle form, later forming an ice mush.

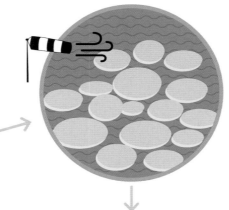

When the wind and waves are strong, "pancake ice" forms: round ice floes that are slightly bent up at the edges due to the movement of the waves. In the next step, the ice floes form a thin sheet of ice that is pushed by the wind towards the existing pack ice.

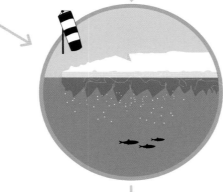

When there is no wind, a thin sheet of ice forms from the ice mush, the "nilas." During freezing, the salt contained in the seawater is deposited into channels in the ice. Over months and sometimes years, it escapes from the underside of the ice sheet in the form of brine, which is why older sea ice consists of fresh water.

1 m
0 m

−9 m

In the first year, the ice cover grows up to 90 cm. Wind and ocean currents push the ice over each other, forming rugged surfaces. After some time, 10 to 20 m thick pack ice ridges form.

In the Arctic, there is perennial 4–5 m thick ice, which survives warmer summers and grows to twice the area in winter.

1 year

In Antarctica, the sea ice cover is annual, meaning that it melts seasonally. In summer, 80% of the ice disappears, and in winter the ice usually grows again.

Melting sea ice...

hardly changes the sea level.

Sources: WOR (2019), AWI (2015)

Ice from the Land

Freshwater

Glaciers consist of fresh water and are constantly in motion. The ice flows slowly towards the valley, forming glacial lakes and feeding rivers. Large pieces of ice break off at the glacier front and then float on the seas as icebergs of various colors and shapes.

Ice has a lower density than water, so it floats. Ocean currents and winds set the direction, and the sea temperature determines the lifetime of the floating ice, which can range from days to years.

Snow-white ice

Sharpened

Transparent ice appears black

Dome

Grey ice, covered with stones and dust

Wedge-shaped

Colored ice with algae growth

Block-like

Blue ice

Table iceberg

Greyish ice with soot particles

U-shaped (dry dock)

Iceberg coloring depends on the type of ice, age, and amount of oxygen, dirt, inclusions, fouling, and last snowfall.

When the sun shines on the ice, the radiation of long wavelengths (red) is absorbed, hence the bluish shimmer of the ice.

Melting glacier ice...

contributes to sea level rise.

North Pole...

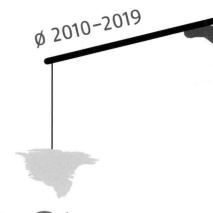

Ø 2010–2019

3200 m

thickness of the Greenland Ice Sheet

243 Gt

of ice mass is lost from
the Greenland Ice Sheet
per year on average.

Polar bears

only live in the Northern Hemisphere.

−18 °C

the annual average
temperature in the
Arctic (on land).

3 M km³

Ice volume in Greenland

8 states

are Arctic neighbours. The territorial claims are governed by the
law of the sea, but are disputed.

Up to **20 m**

mighty pack ice ridges form in
the Arctic sea ice due to wind
and currents – an obstacle that
even icebreakers cannot
overcome.

 **4 M**

people live in the Arctic.

Sources: AWI (2020), IPCC (2021), WOR (2019)

...vs. South Pole

148 Gt

of ice mass is lost
from the Antarctic
Ice Sheet per year.

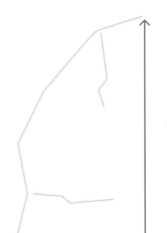

4897 m

thickness of the
Antarctic Ice Sheet

−50 °C

average
temperature at
the South Pole.

Penguins

only live in the Southern
Hemisphere.

26 M km³

Ice volume in the Antarctic

Southern Ocean

Antarctica is surrounded by the sea.
The Antarctic Treaty determines the
peaceful use of the South Pole.

327 km/h

was the wind record measured in
1972 at the Dumont–d'Urville
station in West Antarctica.

1000−4000

researchers live temporarily in Antarctica.

Ice is essential

for the climate and living creatures...

Temperature regulator
Bright ice surfaces reflect a large part of the sun's rays and are therefore a significant factor in the global climate.

Habitat
Countless birds, marine animals, and land mammals in the polar regions depend on snow and ice, for example, to rear their offspring or to hibernate.

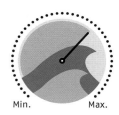

Ocean regulator
Melting glacial ice raises the sea level, changes the salinity and density of the water through the melting process, and even influences ocean currents.

Helper in the search for food
Polar bears and other predators depend on the sea ice for food. Some of the 4 million people living in the Arctic also hunt on the ice.

Water supplier
Almost 69% of the Earth's freshwater is stored in glaciers, and only 30% in groundwater.

Soil protector
In the polar regions, the permanently frozen ground (permafrost) protects against landslides, soil sinking, and erosion* of coasts.

*natural abrasion of rock and soil by water, ice, and wind

...and Humans

are dependent on ice.

Sea level
Rising temperatures are causing more ice sheets to melt, resulting in rising sea levels and flooding. This already poses a serious threat to around 700 million people on coasts, rivers, and islands.

Agriculture
In many regions of the world, such as the Himalayan Mountains or Chile, the cultivation of fruit, vegetables, and cereals in the dry season is only possible thanks to melting water from glaciers.

1350 million

people in high mountain regions and low-lying coastal zones are directly dependent on the intact state of the cryosphere.

Water quality & health
Heavy metals and chemicals such as CFCs stored in the ice can contaminate drinking water as ice melts.

Living expenses
From jobs in shipping or construction to tourism, millions of employees rely on the condition of the ice.

Infrastructure and houses
Infrastructure built on permafrost and houses for 4 million people in the Arctic will only remain stable if the ground underneath is permanently frozen.

Culture
For the Indigenous peoples of the Arctic, ice is particularly identity-forming.

Tourism and leisure
Skiing and other winter sports, ice climbing, or dog sled races on frozen lakes have one thing in common: they need ice and snow.

Sources: IPCC (2019), Wang et al. (2019), WOR (2019)

The "Eternal" Ice in Figures

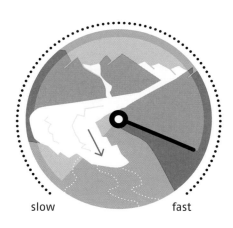

slow   fast

110 m

In 1953, the Kutiah-Lungma glacier in Pakistan moved 110 m downhill per day for a short time, more than 12 km in three months. This made it the fastest glacier in the world. Normally, movements are measured in meters per year.

1967   2020

There is about
53%
less snow in the Northern Hemisphere today compared to 1967.

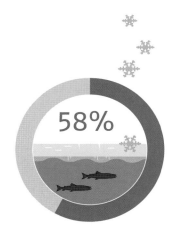

58%

of rivers and lakes in the Northern Hemisphere freeze seasonally.

99%
of the Earth's ice masses are in the ice sheets.

1%
in sea ice, permafrost, freshwater ice, and in the atmosphere.

There are about
220,000
glaciers in the world. They range from sea level (Alaska) to over 8000 m (Himalayas). Glaciers around the world are currently losing more ice mass in summer than they can regain in winter.

For **1450** years, the Arctic Ocean has not been as warm as it is today. The average annual temperature in the Arctic is already 2 degrees warmer.

Scientists can look into climate history for up to **800,000** years using ice cores. Based on snow deposits, particles, and gas bubbles trapped in the ice, as well as its age, scientists can reconstruct features of the past climate, such as CO_2 levels.

3 km

The data is then fed into complex computer models, compared with today's measurements, and used to calculate future forecasts.

Approx. **1300 billion t** of carbon is estimated to be stored in the Arctic permafrost...

...about twice as much as is currently in the atmosphere.

Ice core drilling

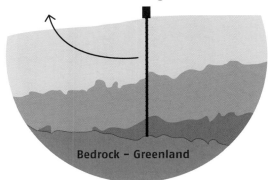

Bedrock – Greenland

Approx. **23** million km² of ground is permanently frozen in the Arctic. Depending on the reduction of CO_2 emissions, about 25–75% of the near-surface permafrost will be thawing by 2100.

Sources: GCW (2021), IPCC (2013, 2019, 2021), NSIDC (2021), Slater et al. (2021), WOR (2019)

Melting Ice, Rising Water Levels

Sea level rise has accelerated in recent decades, from 3.25mm per year to 3.70mm per year. 35% of the rise is now due to melting ice sheets.

Where does the water come from?

	1993–2018	2006–2018
Antarctic Ice Sheet	9.0%	13.1%
		22.3%
Greenland Ice Sheet	15.2%	14.8%
Water storage on land	10.8%	15.4%
Glaciers	19.3%	
Thermal expansion	45.7%	34.4%

1993–2018
Ø 3,25 mm per year

2006–2018
Ø 3,70 mm per year

Heat energy absorption
(2010–2018)

Atmosphere 2

Continents 5

3 Melting ice

90%
Oceans

The oceans absorb the greatest amount of excess heat from the atmosphere. As the water warms, it expands.

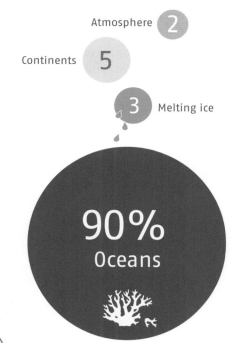

Sources: IPCC (2021), Schuckmann et al. (2020)

Sea level changes 2081–2100 ∎ –1.5 to –1.0 m ∎ –0.2 to 0 ▫ 0 to 0.4 ∎ 0.4 to 0.6 ∎ 0.6 to 0.8 ▸ > 0.8
(relative to 1986–2000)

Future sea level

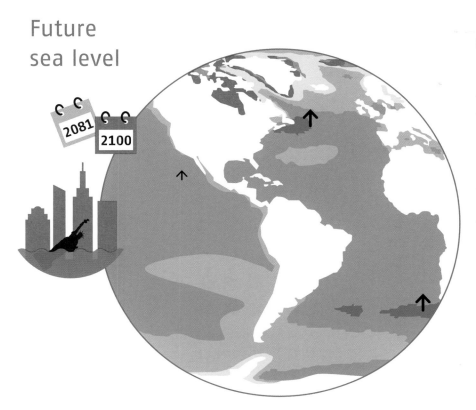

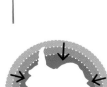
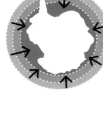

Sea level varies regionally by +/–30%. This
is due, among other
things, to changes in
the prevailing winds,
high and low-pressure
areas, climate pheno-
mena such as El Niño,
ocean currents, and
also the gravitational
forces of the two large
ice sheets.

Contribution to sea level
if the ice sheets were to
melt completely:

Greenland
7 m

all remaining
ice masses: 0.32 m

Antarctica
57 m

The major uncertainty in sea
level rise projections is that we
do not know exactly how fast
the ice masses in Antarctica
and Greenland are melting.

Projection with
faster collapse of
the ice sheets

IPCC scenario
"Worst Case"
(SSP5)

IPCC scenario
"Best Case"
(SSP1)

240
220
200
180
160
140
120
100
80
60
40
20
0

Projections of sea level rise in cm (until the year 2100)

There are five climate scenarios from the Intergovernmental Panel on Climate Change (IPCC), known as 'shared
socio-economic pathways' (SSPs). They predict sea level rise
under different scenarios. In the best-case scenario (SSP1),
we halve our emissions by 2041, reach net-zero by 2050, and
remove emissions from the atmosphere until 2100. This would
limit global warming to 1.4 degrees by 2100. In the worst case
(SSP5), we continue to rely on fossil fuels and emissions triple
by 2100, leading to a temperature increase of 4.4 degrees.

Sources: ICCP (2019), IPCC (2021), Siegert et al. (2020), Slater et al. (2021)

Water Crisis

Less Water for More People

(1) Glaciers and ground water reserves are shrinking...

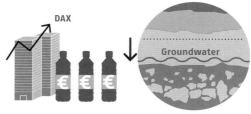

Mineral water companies are lowering the groundwater level in some regions by pumping out too much water. As a result, the population becomes dependent on expensive bottled water.

Droughts and rain redistribution are being exacerbated by the climate crisis.

Rapid population growth, coupled with the man-made climate crisis, is causing severe water shortages. In countries where water scarcity is already severe, it will be exacerbated by the climate crisis. Through changing rainfall patterns and drier climates, it will also spread to new regions of the world. Weather extremes and the melting of inland glaciers are expected to further intensify the problem.

Currently, around 4 billion people are seasonally affected by acute water scarcity, and 500 million people suffer from water shortages throughout the year.

To preserve water as a resource for future generations, we need to better protect water resources.

First and foremost are more efficient agricultural irrigation systems, better use of rainwater, and less water consumption in the production of goods and energy.

Once glaciers have completely melted, water levels in rivers and lakes will shrink.

Sources: Various s. p. 210

② Water consumption is increasing...

Energy demand is growing: large nuclear power plants require over 3 billion liters of water per day.

Industrialization is growing: water-intensive processes include the extraction of raw materials and the processing of rare earths.

Consumption is increasing: It takes around 2500 liters of water to produce one cotton T-shirt.

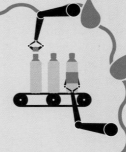

Production of goods: It takes twice as much water to make a plastic water bottle as there is in it.

Inefficient irrigation and increasing demand due to the climate crisis: 70% of the world's water is used in agriculture.

Growing meat consumption: Around 15,500 liters of water are needed for 1 kg of beef.

③ The climate crisis is worsening the situation.

+1/3

More drinking water will be needed in 2050. Drivers include the steadily growing world population, the economy, and the advancing industrialization of emerging countries.

Water scarcity projections for 2040 (less water in %):
- medium to high (20–40%)
- high (40–80%)
- extremely high (>80%)

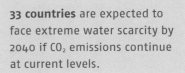

33 countries are expected to face extreme water scarcity by 2040 if CO_2 emissions continue at current levels.

Life-saving Meltwater

Snow is an important reservoir of fresh-water. When it rains in the mountains, the water immediately runs off and seeps away. When it snows, however, water is stored and is available in the spring as meltwater.

Earlier snowmelt caused by climate change is threatening this vital reservoir of water for people, animals, and plants. In some regions, there is now so little snowfall that there is little or no meltwater in spring. This is putting ecosystems and the people who have adapted to them at risk.

Extreme snowstorms become more frequent and threaten people when avalanches bury houses and masses of meltwater flood villages, eroding the soil.

CO₂ domino effect
In the Rocky Mountain watershed, scientists have found that trees can absorb less CO_2 in spring if the snow melts too early. Meltwater reaches the roots too soon, when the trees are still in "winter dormancy." The trees are unable to use the meltwater at this time of year and are deprived of it as spring progresses.

1 billion people depend on meltwater worldwide.

Snow is a reservoir of fresh water

In most regions of the Northern Hemisphere, less snow falls in winter than 50 years ago. Shown here: the USA, for which there is consistent data since 1950.

Chart:

- y-axis: % of precipitation falling as snow (0%, 20%, 40%, 60%, 80%)
- x-axis: 1950, 1980, 2010 2015
- Labels: More snow, More rain, Tendency

Sources: Various see p. 210

Glaciers in the US

1916

2016

Andrews Glacier,
Rocky Mountain National Park,
Colorado

The volume of glaciers in the American West has decreased by about 40% compared to the 1950s in the face of a warming climate.

Check in on your local glacier!
Find glacier photos and the location of glaciers in this interactive map:
glims.org/maps/glims

5 m

3

1

Snowfall
Snowfall is declining in general across the US, and the period during which snow remains frozen is getting shorter.

1 year

Altitude
Perennial snowfields are often found at elevations above 3,000 meters in the Rocky Mountain NP. Below this elevation, no new ice masses can form.

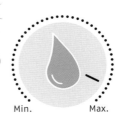

Min.   Max.

Glacier-fed rivers
can carry 30% of meltwater for a short time due to snowmelt.

High water level in spring/ early summer

Meltwater from the glaciers acts as a "buffer"

Low water level in late summer

Water availability
As long as large glacier masses exist, they provide additional water through ice melt during summer for plants and animals, and for use in cities and farmland.

Early melt
In spring, the snow starts to melt 2–3 weeks earlier in the Rocky Mountain NP, with consequences for the enviroment and biodiversity.

Sources: Fountain (2018), McGrath (2019), NPS (2018 & 2023)

Water Straight from the Tap...

Direct indoor water consumption

in the US, per household/quarterly

💧 = 440 liters

24% **Toilet**	🚽	💧💧💧💧💧💧💧💧💧💧💧💧💧💧💧💧💧💧💧💧💧💧💧💧💧💧	11,280 L
23% **Baths and showers**	🚿	💧💧💧💧💧💧💧💧💧💧💧💧💧💧💧💧💧💧💧💧💧💧💧💧💧	10,800 L
20% **Faucet/dishwasher**	🍽️	💧💧💧💧💧💧💧💧💧💧💧💧💧💧💧💧💧💧💧💧💧	9500 L
17% **Laundry**	🧺	💧💧💧💧💧💧💧💧💧💧💧💧💧💧💧💧💧💧	7730 L
12% **Leaks**		💧💧💧💧💧💧💧💧💧💧💧💧💧	5790 L
4% **Other**		💧💧💧💧	1800 L

Total per quarter
and household: **46,890 liters**

That's enough water to fill a small swimming pool, every quarter:

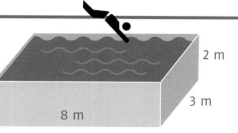

2 m

3 m

8 m

= 521 per day

Sources: DeOreo et al. (2016). Mekonnen & Hoekstra (2011, 2012). Cazcarro et al. (2022). Schyns et al. (2017)

... and Indirectly Consumed Water

Indirect water consumption

for the production of 1 kg (global ø)

💧 = 440 liters

1 kg of beef consumed every month
= 184,800 L of water consumed annually

Per 1 kg (conventional farming)

15,400 L	Beef
10,100 L	Chickpeas
5875 L	Lentils
4325 L	Chicken
2480 L	Rice
1850 L	Noodles

ⓘ Water quantities vary greatly depending on the country of origin and type of cultivation. Organic farming requires up to 50% less water!
waterfootprint.org

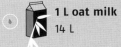 **1 L cow's milk** 940 L

 1 L oat milk 14 L

 1 L wine 436 L

 1 L beer 296 L

 Avocados 1000 L

 Potatoes 287 L

 Apples 822 L

 Bananas 790 L

Rivers of Waste

Industrialized nations vs. developing nations

Tire wear on asphalt in Europe produces an estimated 700,000 million tons of microplastics annually.

Several major cities discharge their wastewater into the sea, including Athens, Barcelona, Brighton, and Cork.

In India, 80% of wastewater is discharged into rivers without any form of treatment.

As many as 700,000 microplastic fibers can be released into wastewater each laundry cycle.

Each year in Europe, about 3800 tons of microplastic from cosmetics and personal care products is released into the environment.

Waste is deposited into illegal dumpsites, where it washes into river systems during heavy rains and floods.

Globally, 500,000 tons of microfibers are released into the ocean from washing machines each year. This accounts for at least 35% of all ocean microplastic pollution.

Illegal gold panning dumps dangerous chemicals such as mercury, a neurotoxin, into rivers in Asia, Africa, and Latin America.

Approximately 7000 different chemicals, including persistent organic pollutants (POPs), are used in textile dyeing processes.

In Europe, it is estimated that 570,000 tons of microplastic granules enter the oceans each year, for example through accidental spillage or through treated industrial wastewater discharged into rivers.

Agricultural fertilizers over-fertilize coastal areas and create oxygen depletion zones, especially near river deltas.

Sources: CSE (2013), UNEP (2021), Boucher & Friot (2017), Westerbos & Dagevos (2022)

9–14 million tons of plastic end up in the oceans every year, which could rise to 23–37 million tons by 2040.

Oceans

Dirty Groundwater in the US

41 %
of 1200 wells tested were contaminated with pesticides from agriculture.

More than
1 in 5
groundwater samples were a concern for human health.

140 million people, nearly 40% of the population, rely on groundwater for drinking water.

Protect your groundwater!

• Buy organic products
(for pollutant-free soils in agriculture).
• Use organic fertilizer in your garden.
• Dispose of paints, batteries, and medicines properly.
• Wash cars only in car washes
(where the water is recycled).
• Use sand instead of road salt in winter.
• Lay water-permeable material on paths and terraces.
• Use a rainwater cistern.
• Water the garden only in the morning or evening.
• Save water in the household.

Sources: Bexfield et al. (2021), DeSimone et al. (2015), USGS (2019)

Groundwater

Priorities in the Global Water Crisis

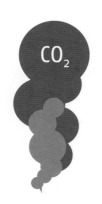

Reduce CO$_2$ emissions ①

According to the Intergovernmental Panel on Climate Change, greenhouse gas emissions must be halved by 2030 (compared to 2010), and by 2050, the world economy must be climate-neutral to keep warming below 1.5 degrees.

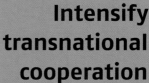

Improve water protection ②

Research into new, water-saving technologies should be pursued to reduce water consumption in industry and households.

Intensify transnational cooperation

with the goal to secure the right to water and a future worth living for generations to come

Use sea & wastewater ③

Wastewater recycling is already being successfully implemented in Singapore, for example. More energy-efficient and cost-effective desalination plants need to be developed.

Irrigate more efficiently ④

More efficient irrigation can increase food security, and protect rivers from low water levels and smallholders from crop failure.

Sources: Circle of Blue (2010), IPCC (2021), UN (2021)

10 Be water wise...  ...and become a water hero!

9 Valuing water

The price of water must be appropriately high to counteract waste and pollution. Human rights must not be violated in the process.

8 Organic farming

Move away from pesticides, biodiversity loss, and industrial monoculture, towards organic farming and diversity, because healthy, humus-rich soils are valuable water reservoirs.

Expand education

Create motivation for new behaviors in consumption and lifestyle. Teach adaptation strategies in the climate and water crisis.

7 Increase renaturation

Look at ecosystems holistically, because everything is interconnected: protecting peatlands, forests, rivers, and their biodiversity is also protecting soil, erosion, water, and the climate.

6 Improve water harvesting

Develop technologies to "harvest" water from the air and capture rainfall.
For countries such as India and Pakistan, this has already become an important climate adaptation strategy.

5 Water use & protection

Multinational companies that produce in developing countries need to be regulated more strictly. They must pay more for their water use to ensure that enough clean drinking water is available for the local population. There must also be tougher penalties for water pollution.

BIOSPHERE

Soils, Plants & Animals

Bio | sphere
[Ancient Greek *bios* = life
& *sphaĩra* = sphere (Earth]
The biosphere is the living part of the
Earth: plants, animals, and humans.
It includes all habitats from forests,
steppes, deserts, and ice to soils, oceans,
and the lowest layer of air.

Soil Life

Underground Ecosystems

Typical soil life

What makes a soil "intact" is its bio-diversity, a complex symbiotic relation-ship between many species. To keep the soil rich in nutrients and organic matter, it needs to be aerated, moist, and fertile.

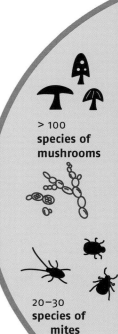

> 100
species of mushrooms

20–30
species of mites

Soils are teeming with life.

Under the micro-scope, billions of organisms can be found in the soil. Bacteria have the largest share.

Soils hold a quarter of the planet's biodiversity. Intact soils are an indis-pensable habitat and a limited resource: once degraded and eroded, it can take centuries or thousands of years for soil to become similarly rich in soil life again.

Preserving diversity and protecting soils is not only important for flora, fauna, and climate but also for the basic needs of us humans. Our food security is at stake.

The EU Biodiversity Strategy for 2030 calls for "more space for nature in our lives." In the climate crisis, the protec-tion, ecological use, and renaturation of soils represent a powerful nature-based solution to capture carbon.

By the way: the well-known compar-ison that "as many organisms can be found on a teaspoon of healthy soil as people on Earth" is not true anymore. It was correct around 1950, but today, one would need closer to a handful.

Save the soils!
Browse information on soil protection, educational material, publications, and graphics:
soils.org/education/classroom

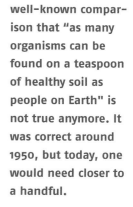

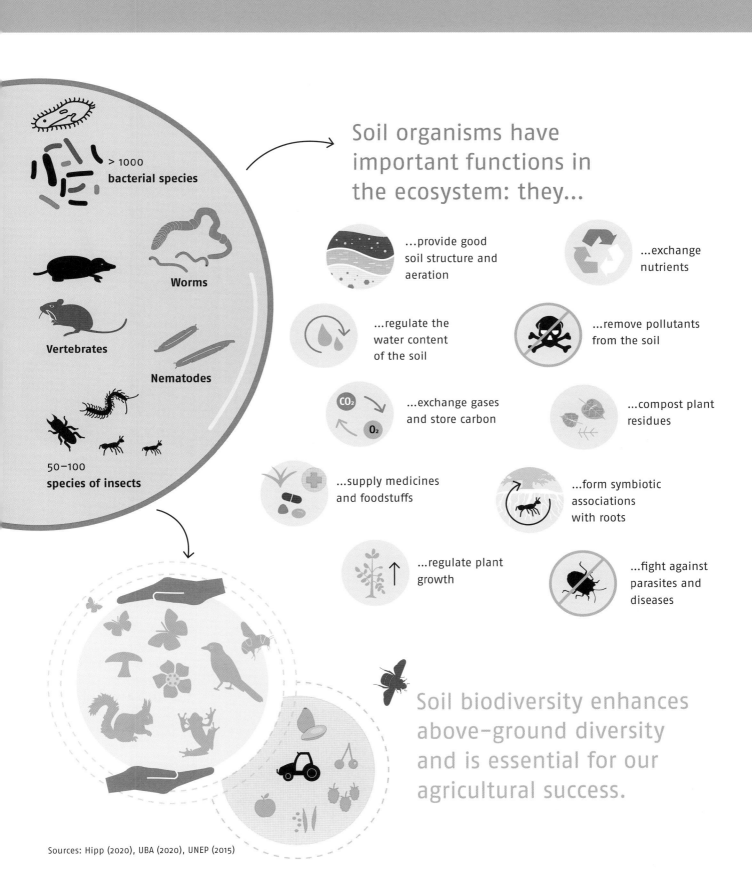

> 1000
bacterial species

Worms

Vertebrates

Nematodes

50–100
species of insects

Soil organisms have important functions in the ecosystem: they...

...provide good soil structure and aeration

...exchange nutrients

...regulate the water content of the soil

...remove pollutants from the soil

...exchange gases and store carbon

...compost plant residues

...supply medicines and foodstuffs

...form symbiotic associations with roots

...regulate plant growth

...fight against parasites and diseases

Soil biodiversity enhances above-ground diversity and is essential for our agricultural success.

Sources: Hipp (2020), UBA (2020), UNEP (2015)

Communication in the Soil

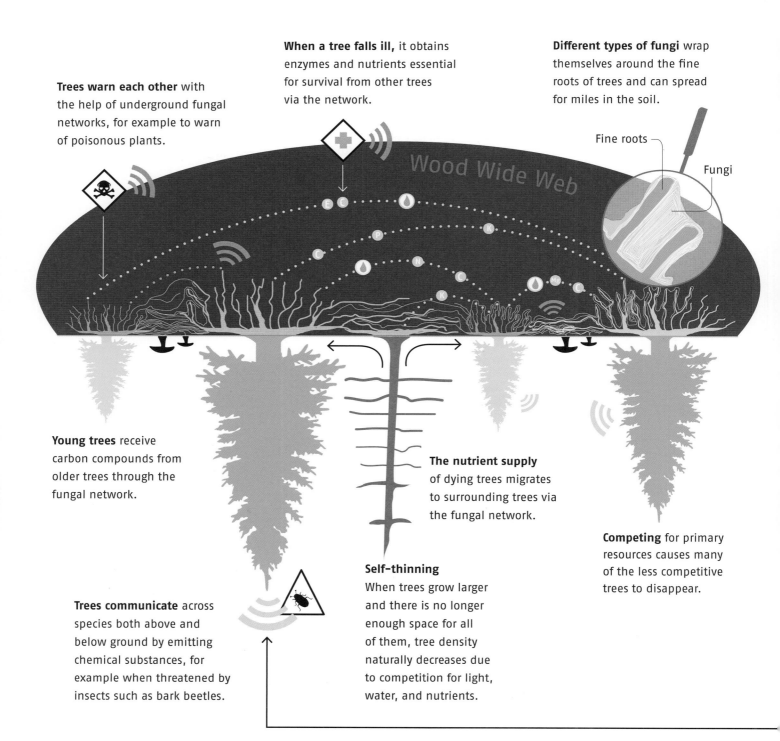

Trees warn each other with the help of underground fungal networks, for example to warn of poisonous plants.

When a tree falls ill, it obtains enzymes and nutrients essential for survival from other trees via the network.

Different types of fungi wrap themselves around the fine roots of trees and can spread for miles in the soil.

Fine roots

Fungi

Wood Wide Web

Young trees receive carbon compounds from older trees through the fungal network.

The nutrient supply of dying trees migrates to surrounding trees via the fungal network.

Competing for primary resources causes many of the less competitive trees to disappear.

Trees communicate across species both above and below ground by emitting chemical substances, for example when threatened by insects such as bark beetles.

Self-thinning
When trees grow larger and there is no longer enough space for all of them, tree density naturally decreases due to competition for light, water, and nutrients.

Water N Nitrogen P Phosphorus K Potassium E Enzymes C Carbon

Older trees are shown dark green and larger, while young trees are lighter colored and smaller.

Wood Wide Web

is the interconnectivity and communication of trees with each other, in symbiosis with fungi and microorganisms. In a patch of forest, the oldest trees are the most intensively interconnected.

Hidden in the forest floor is an extensive network of different types of fungi. These delicate fungi live in symbiosis with multiple trees. They help the tree roots absorb nutrients and water, and in return, they receive sugars from photosynthesis. In times of stress, for example, when a tree does not have enough water or nutrients, the fungal network signals this and substances are exchanged.

Sources: Beiler et al. (2010), Gorzelak et al. (2015), Steidinger et al. (2019)

Soils can Store...

Most carbon is stored in the organic layer (humus). Plants take it up through their leaves and release it back into the soil through their roots. Humus also takes up more carbon through decaying leaves and other organic decomposition processes driven by soil life.

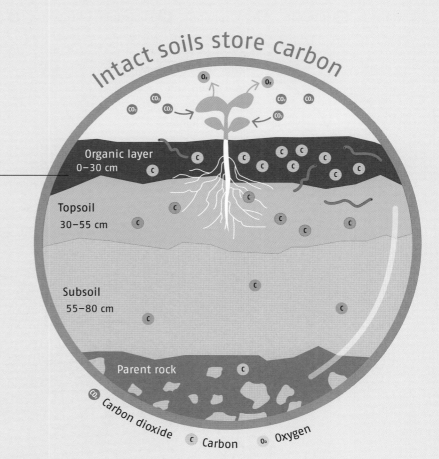

Intact soils store carbon

Organic layer
0–30 cm

Topsoil
30–55 cm

Subsoil
55–80 cm

Parent rock

CO₂ Carbon dioxide   C Carbon   O₂ Oxygen

Carbon storage per ecosystem

Soils are the second largest carbon sink after the ocean. More CO₂ is bound in them than in all of the world's vegetation.

The frontrunner is wetlands such as peatlands, followed by boreal forest soils, but grassland soils also store a lot of carbon.

Carbon in t/ha

200
0
200
400
600

Plants
Soils

Tropical forests
Boreal forests
Temperate forests
Wetlands
Tropical savanna
Temperate grassland
Tundra
Arable land
Desert

Sources: Bazyli & Kryszak (2018), Chemnitz & Weigelt (2015), Flessa et al. (2018), Mukherjee & Kapoor (2018), Patzel & Wilhelm (2018), Wiegandt (2022)

...or Release Carbon

Soil degradation releases CO_2

1 **Excessive use** of machinery, overgrazing, or cultivation of mono-cultures can contribute to the destruction of soil structure.

2 **Over-fertilization** with manure dama-ges biodiversity and soil structure. Fer-tilization can cause erosion as well as humus depletion.

Growing world population

Increasing demand for food

Agricultural overuse

Degradation accelerator

3 **The use of chem-icals** such as herbicides and pesticides kills organisms.

5 **Desertification** and lack of ground cover make the soil more suscepti-ble to wind erosion.

Land clearing through slash and burn

Expansion of agricultural land

4 **Water erosion** after heavy rains damages uncovered, poorly structured arable soils.

In contrast to intact soils, degraded soils release CO_2, are often drier, less fertile and contain fewer nutrients. One-third of all agricultural land is affected by soil degradation, which means that its ecosystem functions are reduced to the point of total loss. Industrial agriculture is responsible for an estimated 25–30% of global greenhouse gas emissions.

Gentle and organic farming practices can regenerate soils and increase car-bon levels, for example through limited plowing, erosion control, green manures, compost and bio-fertilizer. Permaculture rather than monoculture promotes soil quality, as does agroforestry (see p. 196).

25% of the world's soils are degraded.

Underestimated Soils: Grassland

40%

of the Earth's vegetated land area* is grass- land: 52.5 M km².

*excl. Greenland and Antarctica

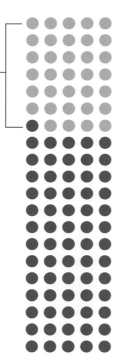

34% of the world's carbon reser- voir is stored in grasslands, of which about 90% is in the soil.

1 billion people world- wide earn part of their income from grasslands.

Grasslands have been part of the Earth's landscape for **20 million** years.

Fire

is a natural component of grasslands and leads to faster growth in spring.

Natural grazing, for example of buffalo herds that are only in one place for a short time and can move on freely, is good for grasslands. The buffaloes "main- tain" the grass, promote greater biodiversity, and eat invasive species such as shrubs.

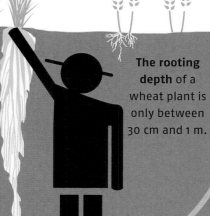

Prairie grass roots are over 2 m long and grow very densely, structuring the soil and protecting against erosion.

The rooting depth of a wheat plant is only between 30 cm and 1 m.

Restoration of the original state

From centuries to "not recoverable"

Destruction and degradation

Depending on the species, from months to decades

(stopwatch dial markings: > 100 000 years · Month · 1000 years · 1 year · 100 years · 10 years)

How grasslands are being rapidly destroyed:

Intensive plowing
as well as use of pesticides and pollutants

Conversion
to monoculture or afforestation

Open pit mining
of raw materials, such as coal or copper

Overgrazing
compacts and erodes the soil.

How to use grassland sustainably:

* incl. clover, lucerne, beans, and many more

Maintain soil structure
through little or no plowing and permaculture cultivation.

Increase biodiversity
by sowing legumes,* making the soil more nitrogenous.

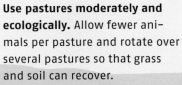

Use pastures moderately and ecologically. Allow fewer animals per pasture and rotate over several pastures so that grass and soil can recover.

In addition, sow legumes and herbs, which contribute positively to biodiversity, root structure, and soil life.

Regulate the pH value
with lime so that the soil does not become acidic and thus unproductive

Promote humus build-up
by planting and incorporating fast-growing plants into the soil.

Sources: Bai et al. (2022), Buisson et al. (2022), Hipp (2020), Isenberg (2000)

Frozen Soils: Permafrost

Soil must be frozen all year round for at least 2 consecutive years (except of the thaw layer) for it to be called "permafrost."

Temperatures

Permafrost only forms above 0 degrees. Temperatures in the frozen ground vary regionally, typically between 0 and −14 degrees. In Antarctica, temperatures as low as −36 degrees were measured.

Patterned surface

When the ground thaws on the surface in summer and freezes again in winter, the freezing process creates 2–50 m polygonal (angular) or circular patterns on the surface.

Ice wedge   Ice lens

Ice accumulation

Trapped ice

The freezing and thawing processes on the surface create crevices and cracks. In summer, the crevices fill with water, which become ice wedges in winter. In addition, oval ice lenses form, which lift the ground and allow ice to accumulate in two-dimensional formations.

Permafrost

- Frozen everywhere (>90%)
- 50–90% of the subsoil frozen
- 10–50% of the subsoil frozen
- submarine permafrost
- perennial ice

Approx. 25% of the land areas in the Northern Hemisphere are permafrost soils.

Vegetation types
Above the frozen ground, a variety of different landscapes form, from tundra and moors to boreal forests, mountains, rivers, and lakes.

Thawing layer
The "active layer" lies above the permanently frozen soils. In summer, this thaws to different depths depending on the region – usually 15–100 cm, in exceptional cases up to 20 m deep.

Bog

Lake

Permafrost

CO_2 CO_2 CH_4 CO_2 CO_2

Gas storage
Permafrost stores large amounts of organic carbon and methane. When released during thawing, they are potent greenhouse gases.

Stones, sediments and plant remains are held together like cement by the frozen water. The amount of ice in the soil varies.

Talik (year-round unfrozen ground)

Expansion
Permafrost extends from a few centimeters to several hundred meters in depth. As a relic of the last ice age, it can reach up to 1.5 km deep into the ground in Siberia.

Sources: AWI (2015), Dobiński (2020), Ehlers (2011), GRID (2020), Overduin et al. (2019), Obu et al. (2019)

Plants

Life's Basic Building Blocks

Approx. **400,000** plant species have been identified worldwide. About 2000 new species are discovered every year.

CO_2

O_2

O_2

Plants produce the oxygen we breathe and form the basis of our food and medicine. We use them for everyday things like clothing (cotton, hemp), building materials (wood, straw), and biofuels (maize, sugar cane). They also have spiritual and cultural significance for many people.

A plant [from Latin *planta* **= seedling, young plant]** is a living organism that grows in the soil, in water, or on other plants. A plant makes its food from sunlight and consists mainly of roots, stems, and leaves. 94% of all plants have flowers.

44% of all plants are threatened with extinction.

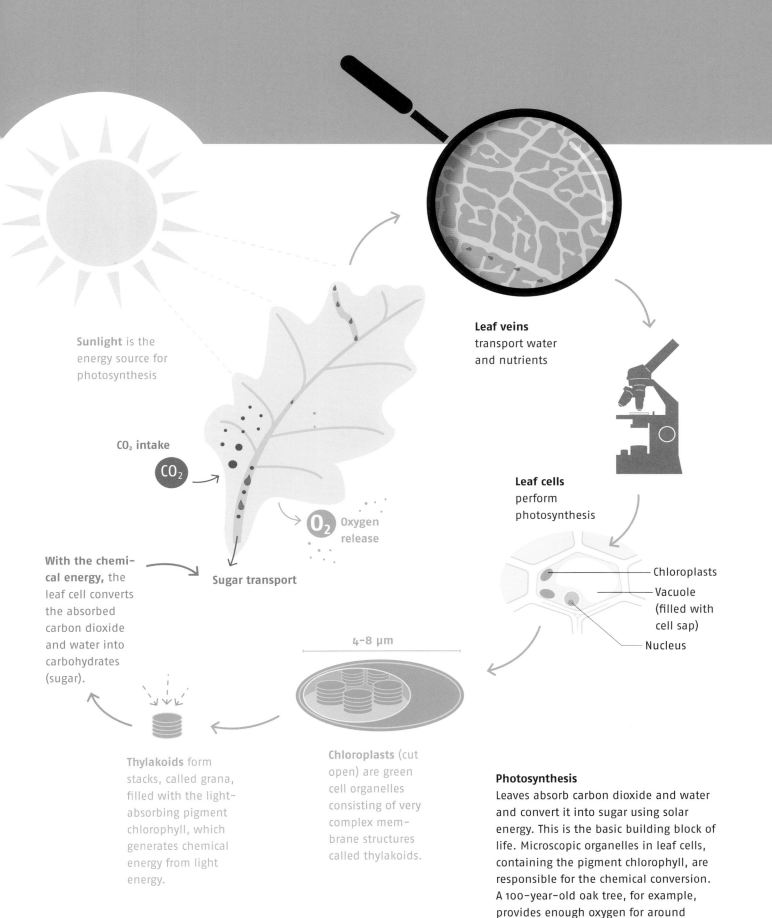

Sunlight is the energy source for photosynthesis

CO₂ intake

CO₂

O₂ Oxygen release

Sugar transport

With the chemical energy, the leaf cell converts the absorbed carbon dioxide and water into carbohydrates (sugar).

Leaf veins
transport water and nutrients

Leaf cells
perform photosynthesis

Chloroplasts

Vacuole (filled with cell sap)

Nucleus

4–8 μm

Thylakoids form stacks, called grana, filled with the light-absorbing pigment chlorophyll, which generates chemical energy from light energy.

Chloroplasts (cut open) are green cell organelles consisting of very complex membrane structures called thylakoids.

Photosynthesis
Leaves absorb carbon dioxide and water and convert it into sugar using solar energy. This is the basic building block of life. Microscopic organelles in leaf cells, containing the pigment chlorophyll, are responsible for the chemical conversion. A 100-year-old oak tree, for example, provides enough oxygen for around 11 people every year.

Sources: Antonelli et al. (2020), BD (2017), Wald und Holz NRW (2022), Spektrum (2001)

The Diversity of Plants

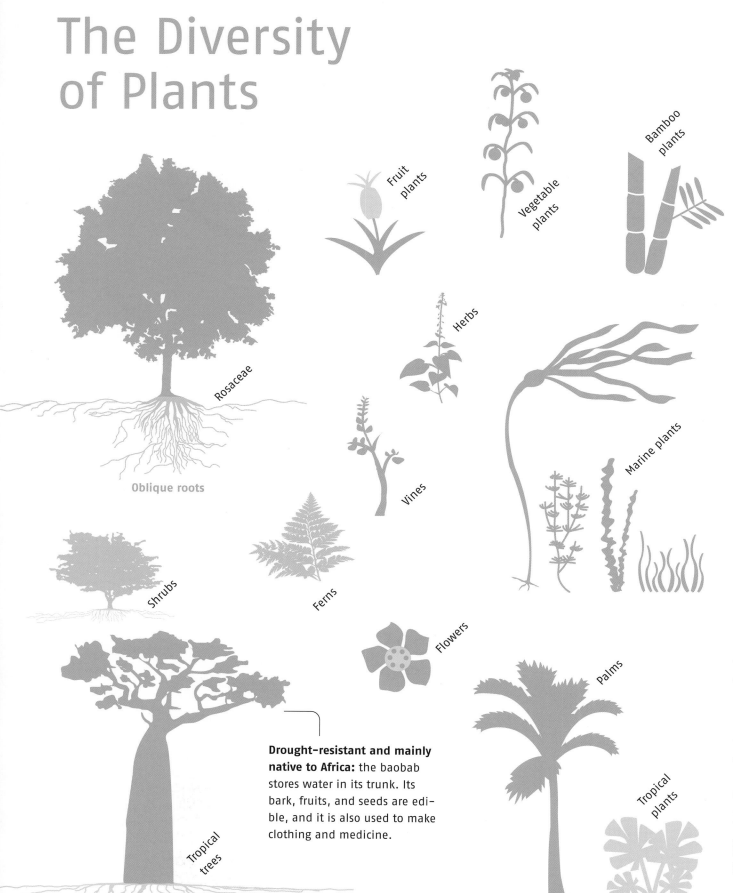

Fruit plants

Vegetable plants

Bamboo plants

Rosaceae

Oblique roots

Herbs

Marine plants

Shrubs

Ferns

Vines

Flowers

Palms

Drought-resistant and mainly native to Africa: the baobab stores water in its trunk. Its bark, fruits, and seeds are edible, and it is also used to make clothing and medicine.

Tropical trees

Tropical plants

Sources: CAL-IPC (2019), Pawlik et al. (2016), Rahul et al. (2015)

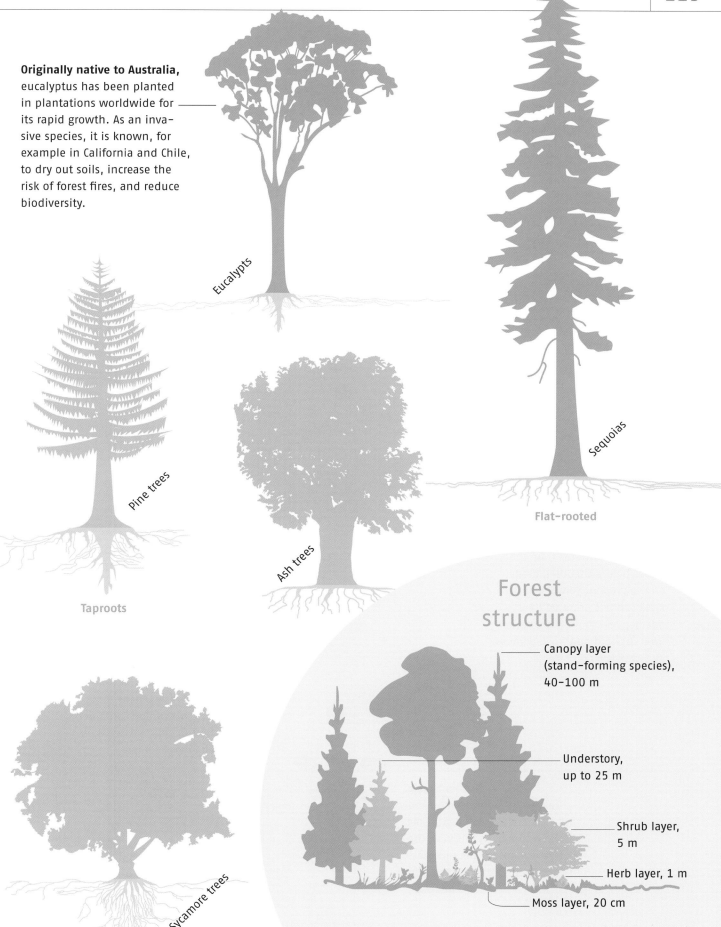

Originally native to Australia, eucalyptus has been planted in plantations worldwide for its rapid growth. As an invasive species, it is known, for example in California and Chile, to dry out soils, increase the risk of forest fires, and reduce biodiversity.

Eucalypts

Sequoias

Flat-rooted

Pine trees

Taproots

Ash trees

Sycamore trees

Forest structure

Canopy layer (stand-forming species), 40–100 m

Understory, up to 25 m

Shrub layer, 5 m

Herb layer, 1 m

Moss layer, 20 cm

The Life of a Tree

Nourishment
Trees live on solar energy and carbon dioxide as well as water, sugar, and dissolved mineral nutrients from the soil. During the growing season, trees need significant amounts of water.

Photosynthesis
see p. 111

CO_2

CO_2

O_2 O_2 O_2

C

C

Winter dormancy
Deciduous trees in cold regions have various mechanisms to protect themselves from the harsh winter conditions. They store sugar as an antifreeze, halt excess water transport and photosynthesis, and extract the remaining nutrients from the leaves before shedding them on the ground. Conversely, most conifers preserve their needles, which are safeguarded from the cold and frost-dryness by a wax layer.

Mg

C
K C
C

Heterotrophic respiration
Microorganisms break down the organic carbon contained in the soil and release CO_2, among other things.

CO_2

CO_2

Mg C K

Mg C K

Nutrient cycle
Leaves are slowly composted with the help of soil organisms such as fungi. In this way, the tree can absorb the nutrients contained in the leaves again via the roots.

C
O_2

CO_2

CO_2

Soil respiration
Part of the carbon is transported from the leaves to the roots, where it is released as CO_2.

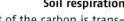
Sources: Beldin & Perakis (2009), Campbell et al. (2003), Leitgeb (2015), Klein & Schulz (2011), SDW (2019)

Water · Oxygen · Nitrogen · Phosphorus · Potassium · Magnesium · Carbon · Carbon dioxide

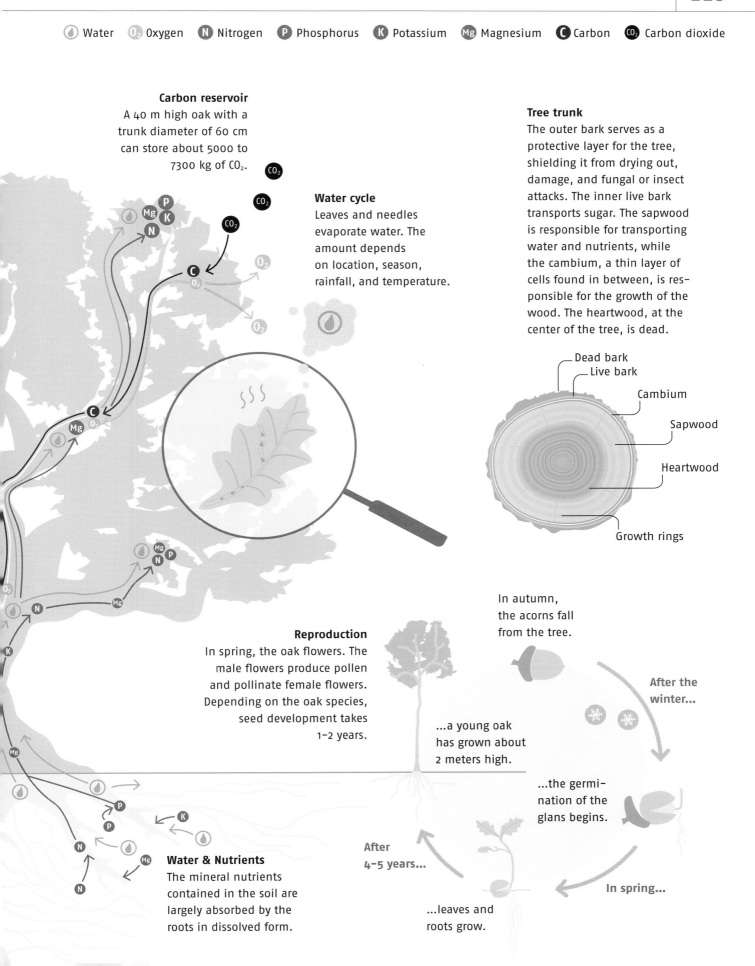

Carbon reservoir
A 40 m high oak with a trunk diameter of 60 cm can store about 5000 to 7300 kg of CO_2.

Water cycle
Leaves and needles evaporate water. The amount depends on location, season, rainfall, and temperature.

Tree trunk
The outer bark serves as a protective layer for the tree, shielding it from drying out, damage, and fungal or insect attacks. The inner live bark transports sugar. The sapwood is responsible for transporting water and nutrients, while the cambium, a thin layer of cells found in between, is responsible for the growth of the wood. The heartwood, at the center of the tree, is dead.

Dead bark
Live bark
Cambium
Sapwood
Heartwood
Growth rings

Reproduction
In spring, the oak flowers. The male flowers produce pollen and pollinate female flowers. Depending on the oak species, seed development takes 1–2 years.

In autumn, the acorns fall from the tree.

After the winter...

...a young oak has grown about 2 meters high.

...the germination of the glans begins.

After 4–5 years...

In spring...

Water & Nutrients
The mineral nutrients contained in the soil are largely absorbed by the roots in dissolved form.

...leaves and roots grow.

The Forests of the Earth

Temperate and subtropical forests

Known for their diversity, these forests range from mixed deciduous and coniferous mountain forests to rainforests. They are mostly found in the Northern Hemisphere in temperate latitudes, but also occur in Chile and Australia. In total, they cover about 27% of the world's forest area.

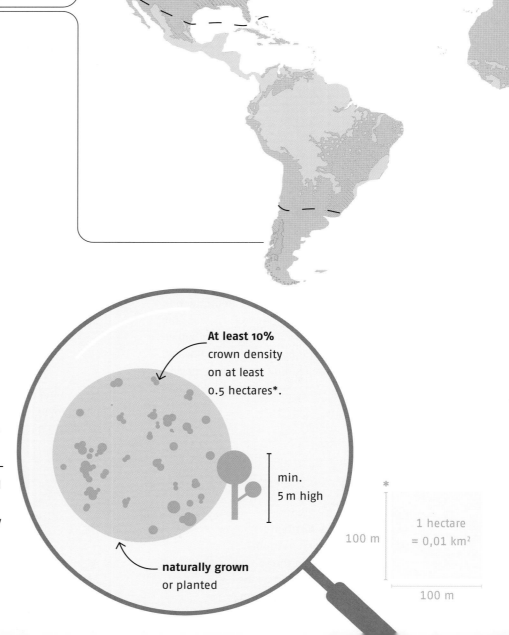

At least 10% crown density on at least 0.5 hectares*.

What is a forest anyway?
There are hundreds of possible definitions. For example, this one from the Food and Agriculture Organization of the United Nations (FAO) defines the minimum size and tree density of a forest:

min. 5 m high

naturally grown or planted

100 m

*
1 hectare = 0,01 km²

100 m

Forest species ■ boreal ■ temperate ▨ subtropical ■ tropical

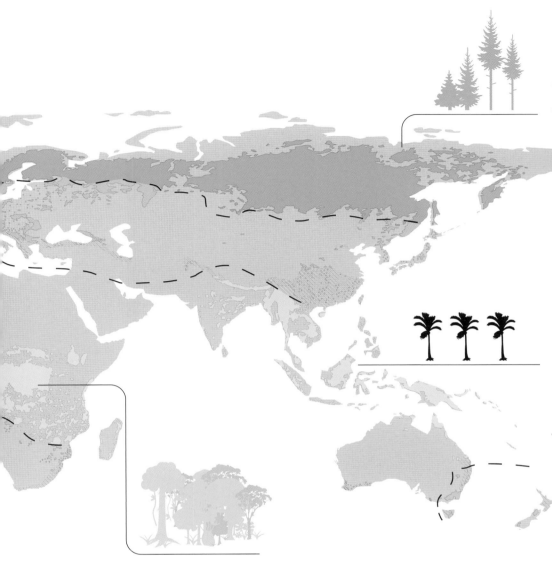

Boreal coniferous forests

The northernmost forests on Earth consist mainly of fir, larch, spruce, and pine. They stretch like a belt across the Northern Hemisphere. As the largest contiguous forest eco-system, they cover about 28% of the world's forest area.

Industrial tree plantations

Between 2010 and 2020, there has been a global expansion of 3 million hectares of new plantations per year. These plantations are predomi-nantly established to meet the increasing demand for wood, pulp, palm oil, or fruits. However, a major problem arises when primary forests are cleared for the purpose of creating new plantations. Since these tree plantations do not fulfill ecosystem functions nor support biodiversity, they are not considered as "forests." According to FAO, they are referred to as "planted forests" and cover only about 3% of the world's forest area.

Tropical forests

With the greatest biodiversity density of all, tropical forests harbor 50% of all known plant species. Besides tropical rain-forests, they include tropical dry forests, mountain forests, and cloud forests. They cover about 45% of the world's forest area and regulate the global climate through processes like water exchange with the atmosphere.

Let's protect old-growth forests!
Roll up your sleeves and get involved, for example with local forest conservation initiatives:
oldgrowthforest.net

Sources: FAO (2020), Gauthier et al. (2015), Jenkins & Schaap (2018), Martone et al. (2017), Petersen et al. (2016), Tyrrell et al. (2012)

Hotspots
of Biodiversity

California
Floristic
Province

North American
Coastal Plain

Mediterranean
Basin

Caribbean
Islands

Madrean
Pine–Oak
Woodlands

Mesoamerica

Guinean Forests
of West Africa

Forests Tumbes–
Chocó–Magdalena

Cerrado

Islands of
Polynesia and Micronesia

Tropical
Andes

Atlantic Forest

Succulent Karoo

Cape Floristic Region

Chilean
Valdivian
Forests

There must be at least **1500** endemic* plant species in
a region for it to be considered a biological hotspot.

* species occurring only in this area

Biodiversity hotspots (colors only to distinguish from adjacent areas)

On average, the 36 hotspots worldwide have already lost **70%** of their original size.

Caucasus

Mountains of Central Asia

Irano-Anatolian

Himalaya

Mountains of Southwest China

Japan

Western Ghats and Sri Lanka

Indo-Burma

Philippines

Horn of Africa

Mountains of Eastern Africa

Wallacea

East Melanesian Islands

Coastal Forests of Eastern Africa

Sundaland

New Caledonia

Madagascar and the Indian Ocean Islands

Forests of East Australia

Maputaland-Pondoland-Albany

Southwest Australia

New Zealand

Stand up for the creation of new nature reserves!
There is not only the IUCN "Red List" of threatened species but also a positive "Green List" for outstanding nature conservation areas: **iucngreenlist.org**

Sources: CEPF (2022), CI (2022), Koenig (2016)

Underwater Forests

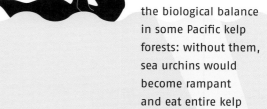

Sea otters establish the biological balance in some Pacific kelp forests: without them, sea urchins would become rampant and eat entire kelp forests bare.

25 % of the coasts are populated by kelp forests.

■ Kelp forests **Threats:** 🦔 Sea urchin desert 🌡 Warming waters

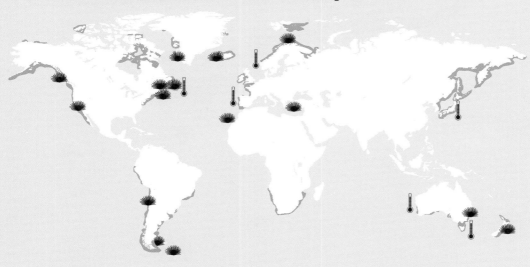

Kelp forests are known as the "forests of the oceans" because of their high biodiversity, productivity, ecosystem dynamics, and similar vertical structure and light conditions.

Kelp forests can play a key role in combating climate change. Not only do they store CO_2 from the environment, but they also help us adapt to the consequences of change: with their higher pH, they are considered a refuge for shellfish threatened by ocean acidification, and in the event of severe storms, they contribute to coastal protection and increase fishermen's yields.

Traditionally cultivated off China and Japan, kelp farming has spread to North America and Northern Europe, with 8 million tonnes harvested globally in 2016. A food rich in vitamins and minerals, it is popular in Asia, where it is processed into food supplements and binding agents.

However, kelp forests around the world are under threat from human impacts such as marine pollution and global warming. As irreplaceable, unique ecosystems, they are in urgent need of greater protection.

Giant kelp grows up to 60 m high and 50–60 cm per day. Its lifespan is up to 7 years.

Sources: Branch (2018), Calloway (2018), Christie et al. (2003), Estes et al. (1998), Filbee-Dexter & Wernberg (2018), Grebe et al. (2019), Krause-Jensen & Duarte (2016), NOAA (2020), Schiel & Foster (2016), Wernberg et al. (2019)

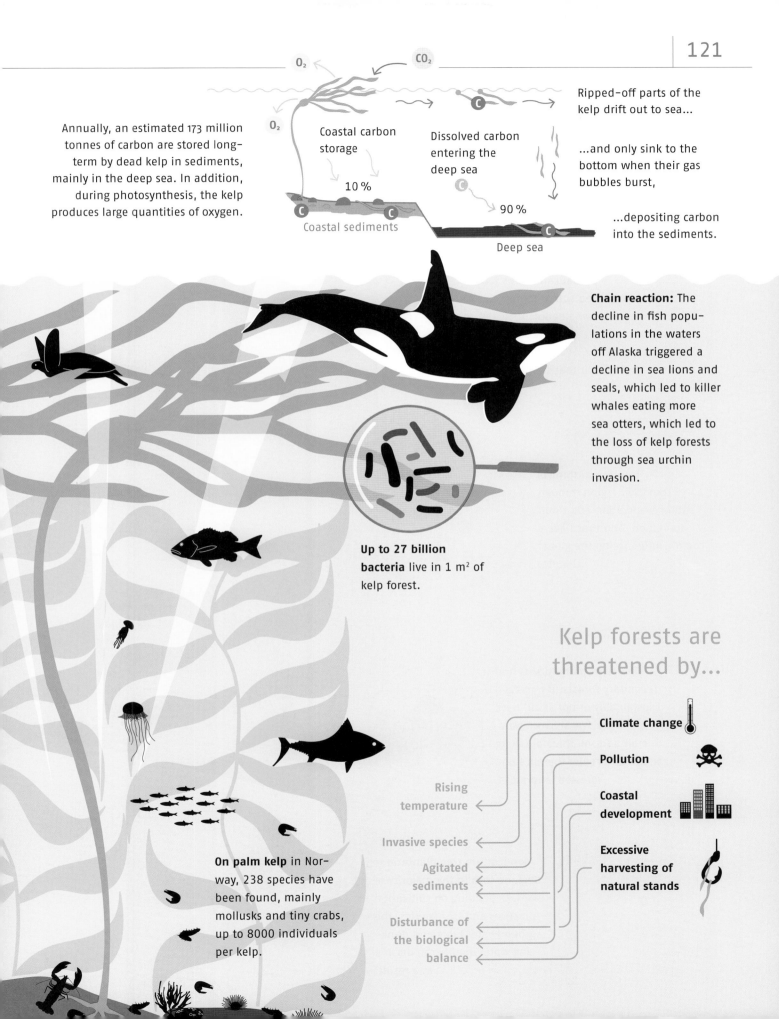

O₂   CO₂

Annually, an estimated 173 million tonnes of carbon are stored long-term by dead kelp in sediments, mainly in the deep sea. In addition, during photosynthesis, the kelp produces large quantities of oxygen.

O₂

Coastal carbon storage

Dissolved carbon entering the deep sea

10 %

Coastal sediments

90 %

Deep sea

Ripped-off parts of the kelp drift out to sea...

...and only sink to the bottom when their gas bubbles burst,

...depositing carbon into the sediments.

Chain reaction: The decline in fish populations in the waters off Alaska triggered a decline in sea lions and seals, which led to killer whales eating more sea otters, which led to the loss of kelp forests through sea urchin invasion.

Up to 27 billion bacteria live in 1 m² of kelp forest.

Kelp forests are threatened by...

Climate change

Pollution

Coastal development

Excessive harvesting of natural stands

Rising temperature

Invasive species

Agitated sediments

Disturbance of the biological balance

On palm kelp in Norway, 238 species have been found, mainly mollusks and tiny crabs, up to 8000 individuals per kelp.

Why Our Climate...

Boreal coniferous forests warm their surroundings: due to their low albedo*, they absorb almost all of the sun's energy (approx. 92 %) and thus warm the air that surrounds them.

Tropical rainforests cool their region through strong evaporation and cloud formation, and thus have a cooling influence on the global climate.

Dense primary forests and secondary forests, if managed sustainably or not at all, have their own microclimate. The air and soil there are more humid and cooler than in non-forested land areas.

Monoculture tree plantations degrade soils in the majority of cases. Depending on cultivation practices and tree species, the groundwater table and drinking water quality can decline as a result.

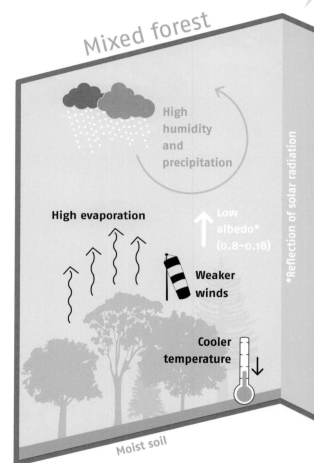

Mixed forest

High humidity and precipitation

High evaporation

Low albedo* (0.8–0.18)

Weaker winds

Cooler temperature

*Reflection of solar radiation

Moist soil

...Depends on Intact Forests

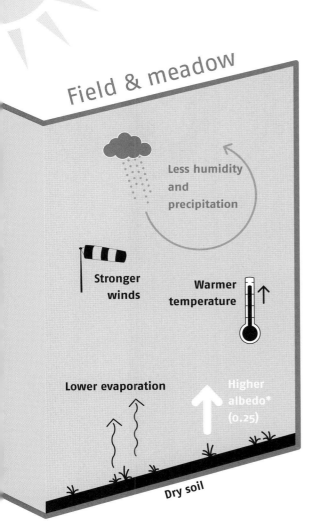

Field & meadow

Less humidity and precipitation

Stronger winds

Warmer temperature

Lower evaporation

Higher albedo* (0.25)

Dry soil

Forests help protect the climate, removing around 28 % of man-made CO_2 emissions from the atmosphere.

Reforestation and renaturation, paired with CO_2 emissions reductions, are an effective climate protection strategy to sequester CO_2 in the long term.

Approximately 296 Gt of CO_2 are stored worldwide in the form of carbon in the above- and below-ground biomass of forests. Further deforestation and degradation would release parts of it.

Sources: Artaxo et al. (2018), Centritto et al. (2011), Sanderson et al. (2012)

The Breath of the Forests

CO₂ sink or source?

Untouched or intact, sustainably managed forests absorb twice as much CO_2 as they emit. This is in contrast to degraded or destroyed forests: after clear-cutting or slash-and-burn, they emit CO_2.

The situation is particularly dramatic in Southeast Asia and the Amazon. The Congo Basin, the world's second largest rainforest, is currently still a CO_2 sink, but it urgently needs to be better protected from (illegal) exploitation.

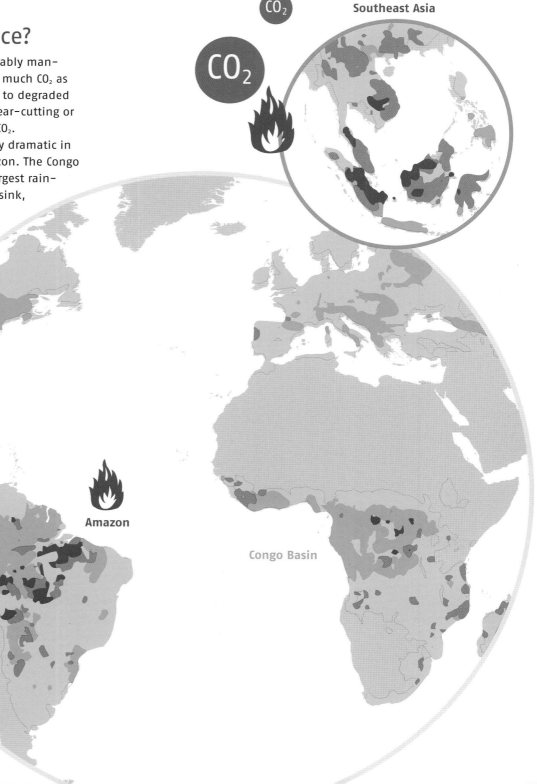

Southeast Asia

Amazon

Congo Basin

Where do trees store the most carbon?

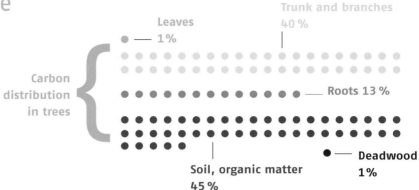

Leaves
1 %

Trunk and branches
40 %

Carbon
distribution
in trees

Roots 13 %

Deadwood
1 %

Soil, organic matter
45 %

During the growing season (from May to October in the Northern Hemisphere), the concentration of CO_2 in the atmosphere decreases. The greener it gets, the more CO_2 is absorbed by the leaves during photosynthesis and stored as carbon, distributed throughout the tree.

A tree in the tropics stores an average of **22.6 kg CO_2 per year.**

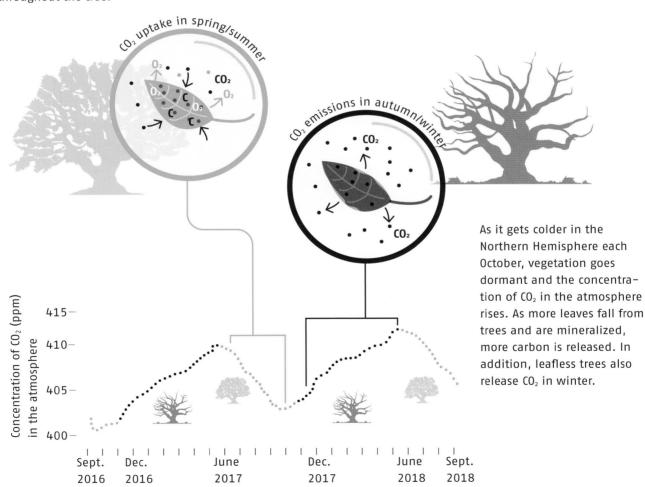

CO_2 uptake in spring/summer

O_2
O_2
CO_2
O_2
C
C
O_2
C

CO_2 emissions in autumn/winter

CO_2
CO_2

As it gets colder in the Northern Hemisphere each October, vegetation goes dormant and the concentration of CO_2 in the atmosphere rises. As more leaves fall from trees and are mineralized, more carbon is released. In addition, leafless trees also release CO_2 in winter.

Concentration of CO_2 (ppm) in the atmosphere

415 —
410 —
405 —
400 —

| Sept. | Dec. | | June | | Dec. | | June | | Sept. |
| 2016 | 2016 | | 2017 | | 2017 | | 2018 | | 2018 |

Sources: Betts et al. (2008), FAO (2018), Harris et al. (2021), IPCC (2014), MPG (2010), Gough et al. (2008), WRI (2018)

Great Green Wall in Africa

18% of the "Great Green Wall," or 17.8 M ha, has been completed.

- Initiator: African Union (AU), 20 states help financially
- Founding Year: 2007
- Length: almost 8000 km
- Width: approx. 15 km
- Targets by 2030:
  - ☐ Restore 100 million ha
  - ☐ Sequester 250 million tonnes of carbon
  - ☐ Create 10 million jobs

The Great Green Wall (GGW) is a mosaic of restoration projects in nearly 30 countries across the Sahel. Land and soil restoration and water improvement are the main focus of the ambitious initiative.

Tree planting is only a small part, as it is very expensive and can only take place once arid conditions have improved.

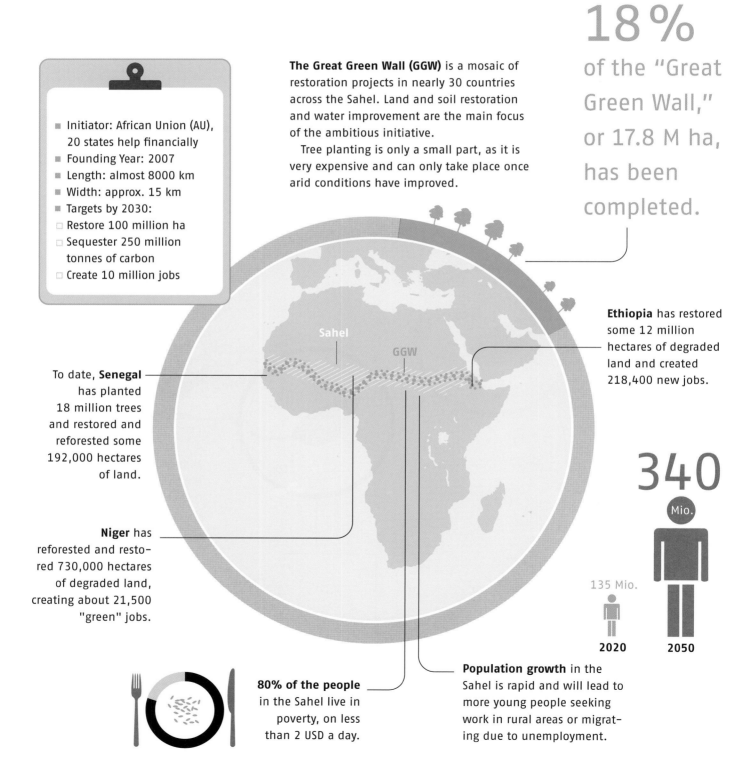

Sahel

GGW

Ethiopia has restored some 12 million hectares of degraded land and created 218,400 new jobs.

To date, **Senegal** has planted 18 million trees and restored and reforested some 192,000 hectares of land.

Niger has reforested and restored 730,000 hectares of degraded land, creating about 21,500 "green" jobs.

340 Mio.

135 Mio.

2020   2050

80% of the people in the Sahel live in poverty, on less than 2 USD a day.

Population growth in the Sahel is rapid and will lead to more young people seeking work in rural areas or migrating due to unemployment.

Renaturation has many goals, depending on the region...

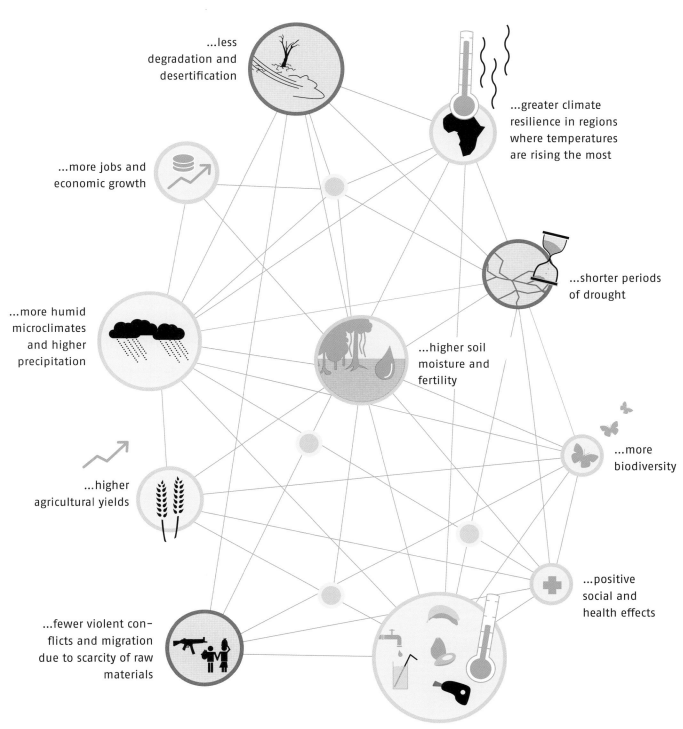

...less degradation and desertification

...greater climate resilience in regions where temperatures are rising the most

...more jobs and economic growth

...shorter periods of drought

...more humid microclimates and higher precipitation

...higher soil moisture and fertility

...more biodiversity

...higher agricultural yields

...positive social and health effects

...fewer violent conflicts and migration due to scarcity of raw materials

...greater food & drinking water security, climate protection, and less poverty overall

Sources: GGW (2020), UNCCD (2020)

When the Climate Becomes Drier...

...insect infestations & diseases increase.

Water scarcity makes conifers, in particular, more susceptible to insect attacks, such as the bark beetle, which benefits from prolonged droughts and has destroyed hundreds of millions of hectares of forest in western Canada and the USA, among other places.

...plants conquer new habitats.

The shift in climate zones also changes the natural mix of species: as conifers spread northwards, deciduous forests expand into the rainier west of the USA, for example.

...forests die.

In California, more than 100 million trees have died after 5 consecutive years of drought, and millions more have been weakened.

...biodiversity declines.

Highly specialized tree species that depend on birds for reproduction, such as the white-stemmed pine (Pinus albicaulis), may be threatened with extinction if the climate changes rapidly.

...plants slow their growth.

When water is scarce, the growth rate decreases, and some plants even stop growing. Leaves are shed prematurely to prevent further water loss.

...the CO_2 cycle changes.

Droughts disrupt processes such as carbon and water cycles in ecosystems. Under stress, tropical forests can become CO_2 emitters instead of sinks.

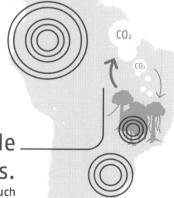

...the soil dries out.

Rivers and groundwater are directly affected by droughts due to reduced precipitation and increased evaporation; water levels and quality decline.

Before the drought

Year 1 drought

>3 years drought

◎ "Flash drought" hotspots: "Flash droughts" dry out the soil within days to weeks.

...droughts intensify worldwide.

Droughts and "flash droughts" are triggered by hotter, drier summers and exacerbated by heavy agricultural and industrial use of groundwater and overpopulation.

...fewer seedlings survive.

In times of drought, the soil moisture is too low for seeds to germinate. With their short roots, seedlings do not reach the moister soil layers and die.

...invasive species spread.

Invasive species can out-compete native species for water, drying out soils. In South Africa, 6–12 years after eucalyptus plantations were planted, the trees spread rapidly and rivers began to dry up.

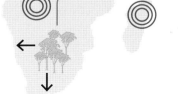

...the risk of forest fires increases.

After 3 years of drought and record temperatures, around 5.8 million hectares of forest burned in Australia from September 2019 to February 2020, 21% of Australia's total forest area.

Sources: BOM (2019), Fei et al. (2017), Forsyth et al. (2004), Kolb (2019), Malhi & Philips (2005), Mukherjee et al. (2022), NCC (2020), Vose et al. (2015)

Halting Species Loss...

Land areas worldwide:

15% protected

At least 35% of the world's land should be protected to help tackle the climate and biodiversity crises.

Ecosystems are made up of a variety of species that live together and depend on each other in a complex network of interrelated organisms.
Scientists call biodiversity our "global safety net." A high variety of life on Earth promotes stronger climate resilience, and provides food and water security. For ecosystems to adapt to the stresses of climate change and to regenerate as quickly as possible after extreme weather events, a high level of bio-diversity is essential.

Many of the United Nations Sustainable Development Goals (SDGs) can only be achieved with restored biodiversity, such as food security, drinking water security, and forest, ocean, and climate protection.If we protect enough natural areas in the right regions, we can still achieve these goals.

Climate change will make this more difficult, however. At more than 1.5 degrees of warming, scientists predict that some ecosystems will collapse as too many species are unable to adapt and become extinct.

Source: Dinerstein (2020)

...and Stabilizing the Climate

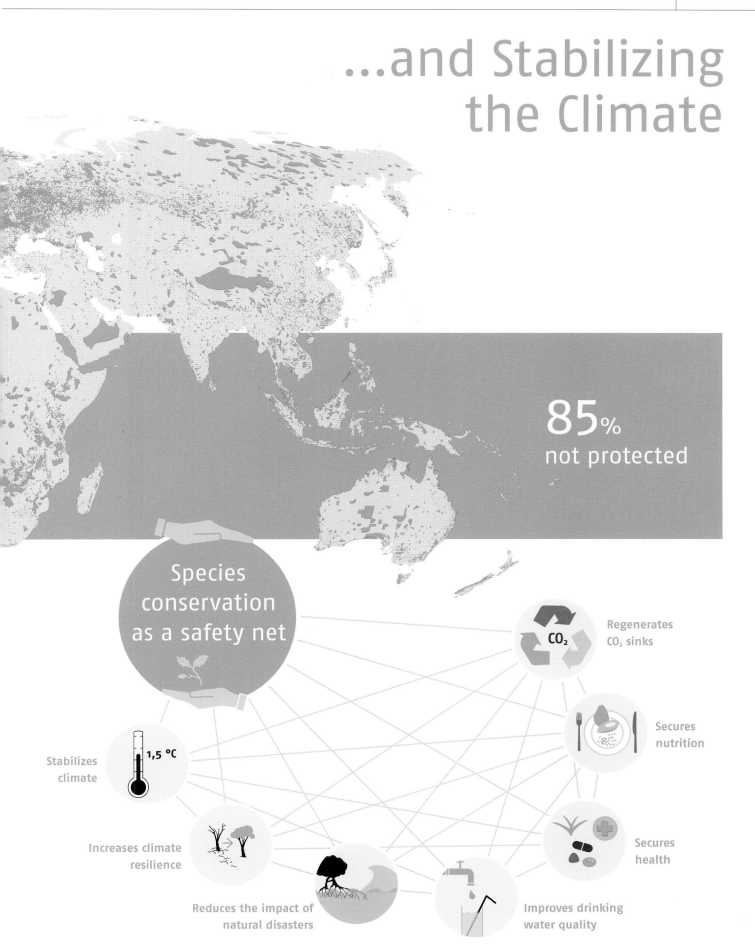

85% not protected

Species conservation as a safety net

Regenerates CO₂ sinks

Secures nutrition

Secures health

Improves drinking water quality

Reduces the impact of natural disasters

Increases climate resilience

Stabilizes climate

1,5 °C

Animals

Global Fauna

Approx. 2.13 M

animal species out of the estimated 8.7 million species worldwide have been identified and named so far.

Taxonomy of animals at a glance

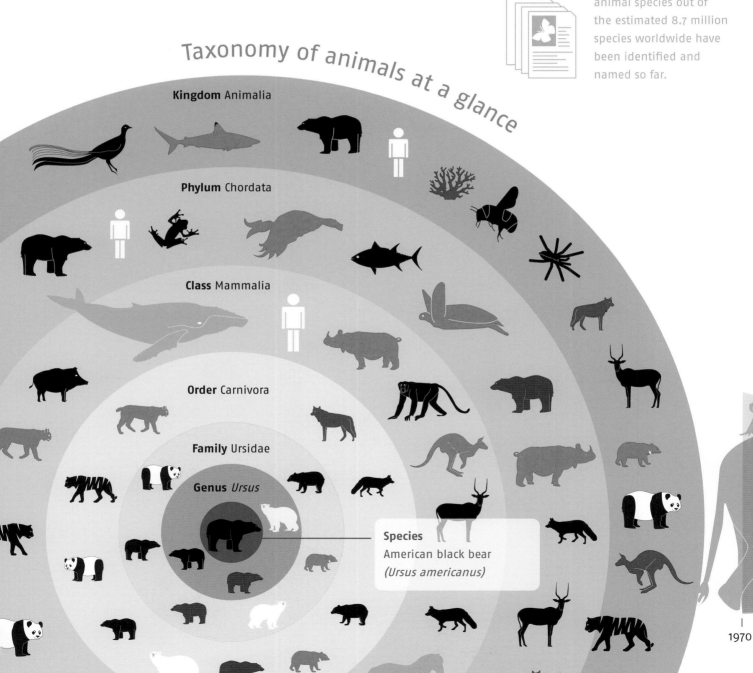

Kingdom Animalia

Phylum Chordata

Class Mammalia

Order Carnivora

Family Ursidae

Genus *Ursus*

Species
American black bear
(*Ursus americanus*)

1970

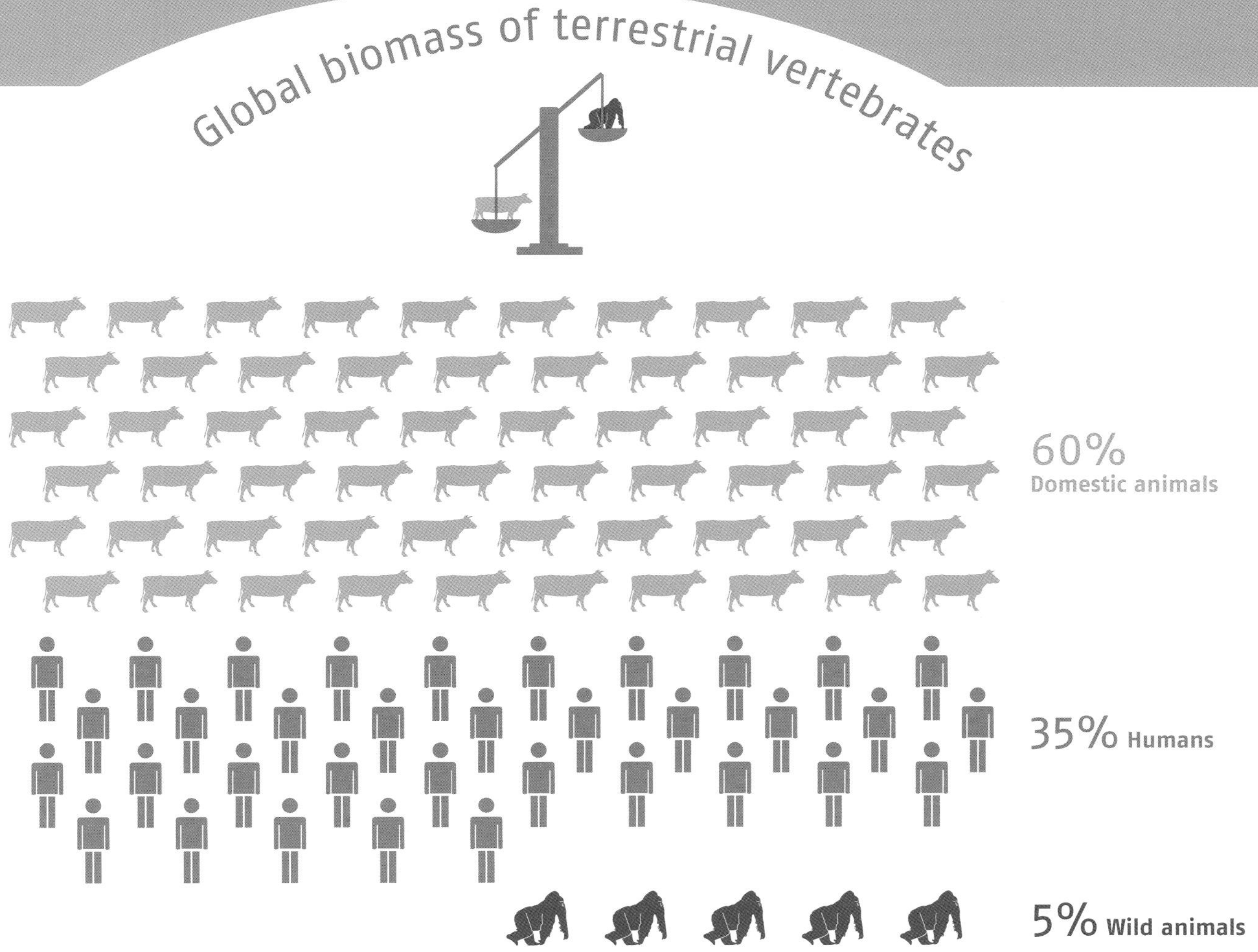

Global biomass of terrestrial vertebrates

60% Domestic animals

35% Humans

5% Wild animals

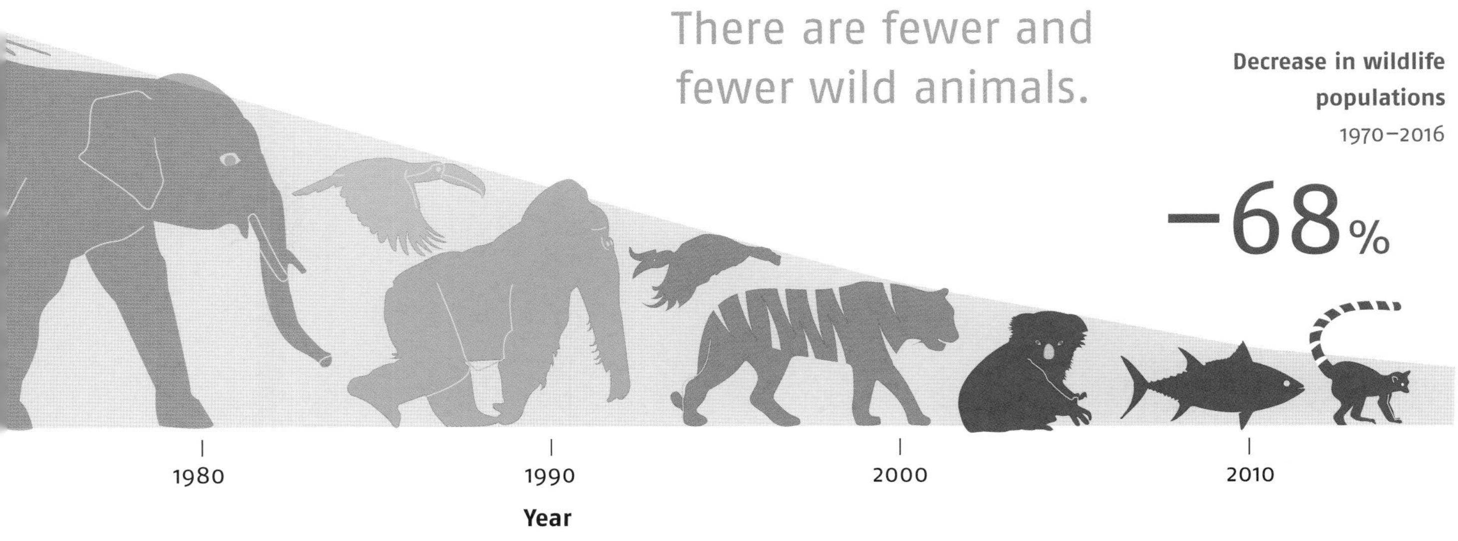

There are fewer and fewer wild animals.

Decrease in wildlife populations
1970–2016

−68%

1980	1990	2000	2010

Year

Sources: IUCN (2021), Mora et al. (2011), Wiegandt et al. (2022)

Complex Food Web

The regenerative capacity of ecosystems is only as strong as their biological diversity. Plant diversity attracts a
wide range of animals, which in turn
play important roles: some provide
pollination, others disperse seeds,
maintain grasslands by eating shrubs,
or fertilize soils.

A food web is an illustration of
the interconnectedness of food chains
in an ecosystem: most organisms are
consumers of, or are consumed by,
more than one species.

**A Northern Hemisphere
forest food web:**

**Carnivores
2nd order**

Lynxes

**Carnivores
1st order**

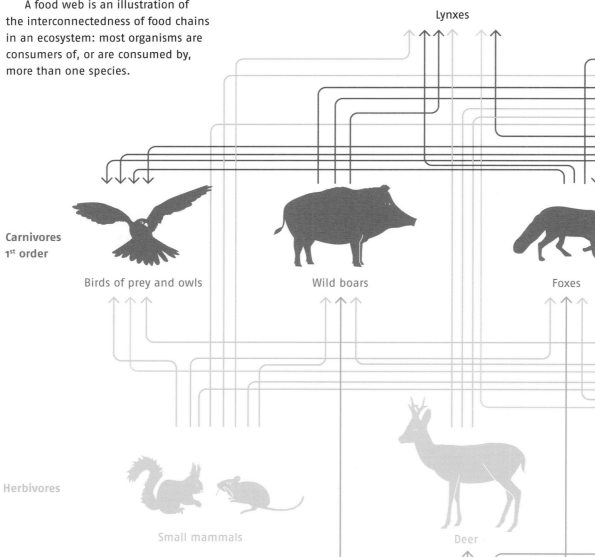

Birds of prey and owls

Wild boars

Foxes

Herbivores

Small mammals

Deer

Plants

Mosses, lichens, leaves, fruits,
seeds, and wild berries

→ The arrows point to the consumer

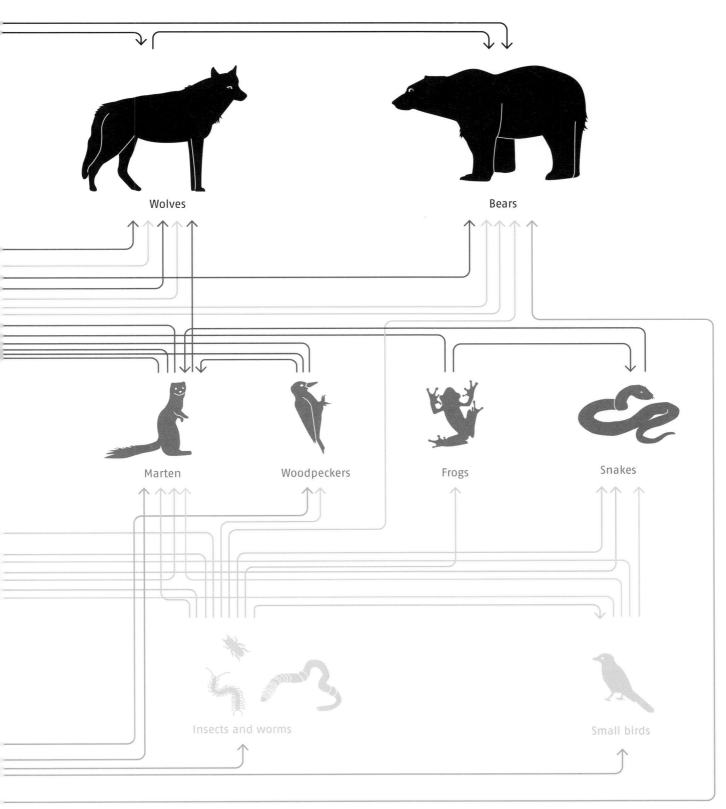

Wolves

Bears

Marten

Woodpeckers

Frogs

Snakes

Insects and worms

Small birds

Loss of Forest Dwellers

Hornbill

Mount glorious torrent frog

Serbian grasshopper

Rusty-spotted bumblebee

Blue ornament bird spider

Giant panda

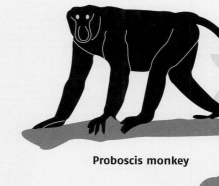

Proboscis monkey

Java rhino

Borneo dwarf elephant

Tiger

Sources: Alroy (2017), IUCN (2020)

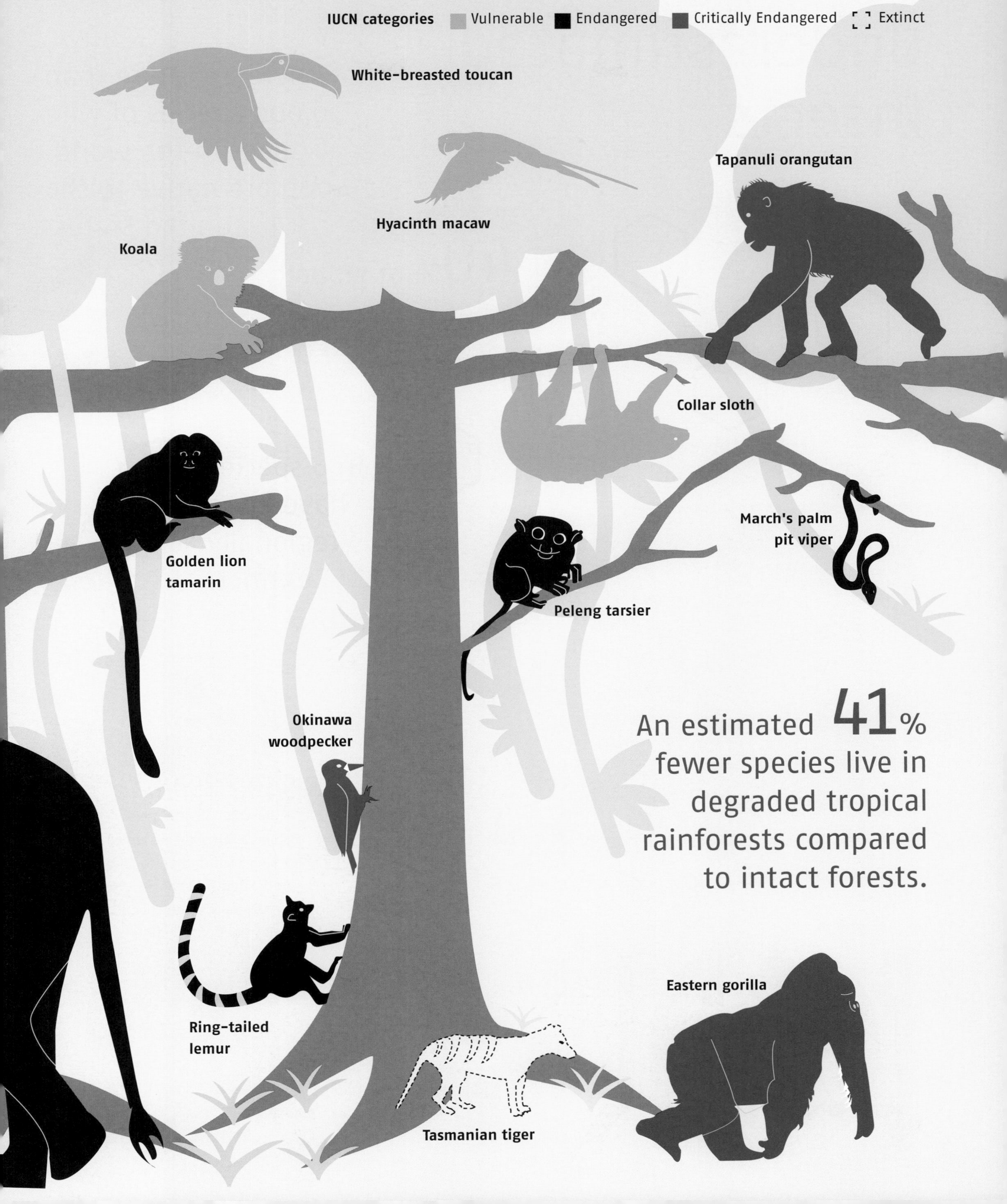
IUCN categories
Vulnerable
Endangered
Critically Endangered
Extinct
White-breasted toucan
Hyacinth macaw
Tapanuli orangutan
Koala
Collar sloth
Golden lion tamarin
March's palm pit viper
Peleng tarsier
Okinawa woodpecker
Ring-tailed lemur
Eastern gorilla
Tasmanian tiger

An estimated **41**% fewer species live in degraded tropical rainforests compared to intact forests.

Underestimated Insects

There are more than 20,000 species of wild bees in the world. 4000 are native to the United States.

40% of global bee species are vulnerable to extinction.

1 in 6 species have gone regionally extinct.

60% of species are stable.

Approx. **3600 USD/ha** is the amount wild bees generate on average, which is their contribution to food production. This makes them slightly more productive than honey bees (3170 USD/ha).

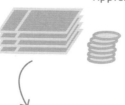

Approx. **35**% of the world's food crops depend on pollinating insects such as bees, wasps, moths, flies and butterflies. Pollinators are essential for the stability of the food supply.

Sources: Bassett & Lamarre (2019), Eggleton (2020), Kleijn (2016), Seibold et al. (2019), USDA (2022), Weber (2019)

Globally, **40**% of all insect species are threatened with extinction.

Main reasons for loss of species

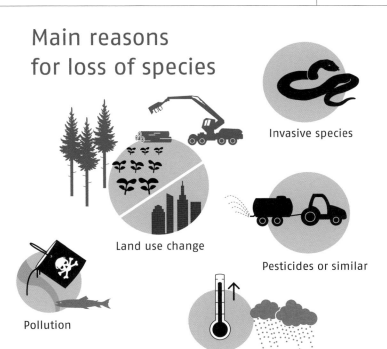

Invasive species

Land use change

Pesticides or similar

Pollution

Climate change

8 steps to help protect insects

1 Increase focus on plant diversity.

2 Reduce and more strongly regulate the use of pesticides and ban neo-nicotinoids.

3 Improve land use, such as allowing "weeds" and wild growth at the edge of fields.

4 Protect and restore peat-lands and natural river courses.

5 Convert mono-culture forest plantations into mixed forests with a higher deadwood content.

6 Cultivate more bee-friendly flower meadows in gardens and parks.

7 Stop the spread of invasive species such as Asian wasps.

8 Fund more research on insects, from identification to climate change impacts.

Icy Habitats

Protect the Arctic!
The Arctic Animal Movement Archive brings together 15 million movement patterns from studies conducted from 1991 to the present day, showing how the Arctic has changed from an animal's perspective:
movebank.org

Acclimatization
Animals try to adapt to new climatic conditions by changing their feeding behavior, metabolism, respiration, blood pressure, or water and nutrient intake.

Migration
Several species of animals on land and in the Arctic Ocean are shifting their habitat towards the North Pole as a result of warming. Polar cod, for example, are increasingly migrating from the Bering Sea towards the North Pole as water temperatures in the Arctic rise by up to 7 degrees Celsius.

> 5 °C

NORTH POLE

New predators
Invasive species are changing food webs. For example, orcas are increasingly hunting narwhals in the ice-free Canadian Arctic, threatening their populations.

Loss of sea ice
Walruses, polar bears, and seals use the sea ice to feed, rest, and raise their young. As the ice retreats, exhausted walruses, for example, are driven to distant beaches because they can no longer find enough ice floes in their hunting grounds.

Health
As a result of the increased effort required to forage, animal health declines and low body weight and malnutrition become more common.

Invasive species
On land, at sea, and in the air, subarctic species are spreading further north. In Alaska, for example, red foxes are already displacing arctic foxes.

Status of the emperor penguin colonies, forecast for 2100
🐧 Critically Endangered   🐧 Endangered   🐧 Vulnerable   🐧 not threatened

Antarctica

Areas with less sea ice on annual average (20th vs. 21st century)
▬ −20%
▬ −16%
▬ −14%

SOUTH POLE

The emperor penguin population is expected to shrink by

50% by 2100.

Sea ice loss
By the end of the century, there will be much less sea ice in Antarctica. But emperor penguins need stable sea ice to nurse their chicks for 8 months.

−60 °C
+37 °C

May

In May, the female lays an egg on the ice. The male then incubates the egg for 2 months while the mother goes out to sea to hunt.

July

About 65 days later, the chick hatches and is kept warm and fed – now by the mother – for another 50 days under her belly feathers.

Sept.

After 2 months, both parents leave to hunt and the "crèche" (group) of penguins gather on the ice to keep warm. Ideally, the sea ice has now reached its maximum extent.

Dec.

The family moves closer to the edge of the sea ice, and the chicks start growing their coat of water-proof feathers and make their first attempts at hunting. The sea ice begins to break up and melt.

Jan. Feb.

The chicks' plumage is now water-resistant and they are ready for long dives and to hunt independently in the deeper ocean. The sea ice should only now reach its lowest extent.

Sources: CBD (2021), Davidson et al. (2020), Drost (2016), Jenouvrier et al. (2014), Klein & Sowls (2015), Lefort et al. (2020), Thyrrin & Sejr (2018), WOR (2019)

On the Rise: Invasive Species

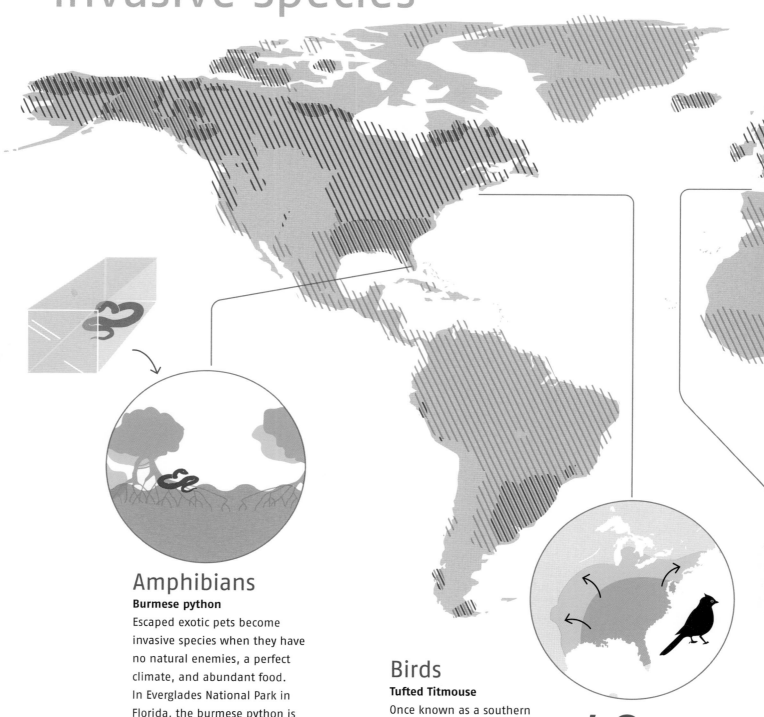

Amphibians
Burmese python
Escaped exotic pets become invasive species when they have no natural enemies, a perfect climate, and abundant food. In Everglades National Park in Florida, the burmese python is responsible for the decline of several mammal species.

Birds
Tufted Titmouse
Once known as a southern species, the tufted titmouse has spread from New Jersey to Maine, 1200 km further north, since the beginning of the 20th century.

48%
range expansion, 1966–2017

Status of invasive species (period 2000–2100) \\\\\\ More invasive species   \\\\\\ Less invasive species   \\\\\\ both

Marine animals
Sea urchin

Tropical herbivores invading temperate latitudes have turned once species–rich kelp forests into sea urchin deserts in Japan, Tasmania, and the east coast of Australia.

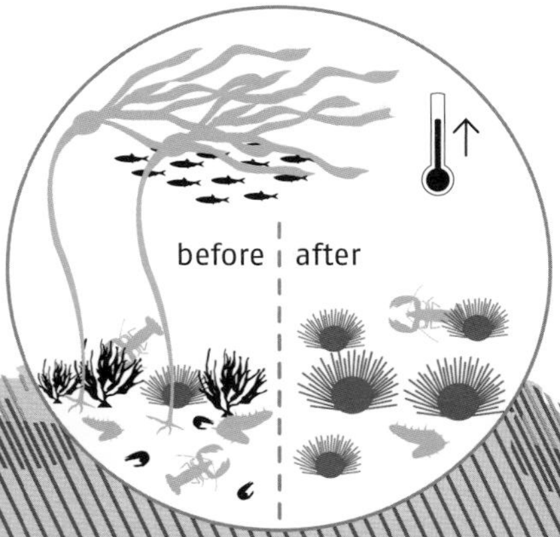

before after

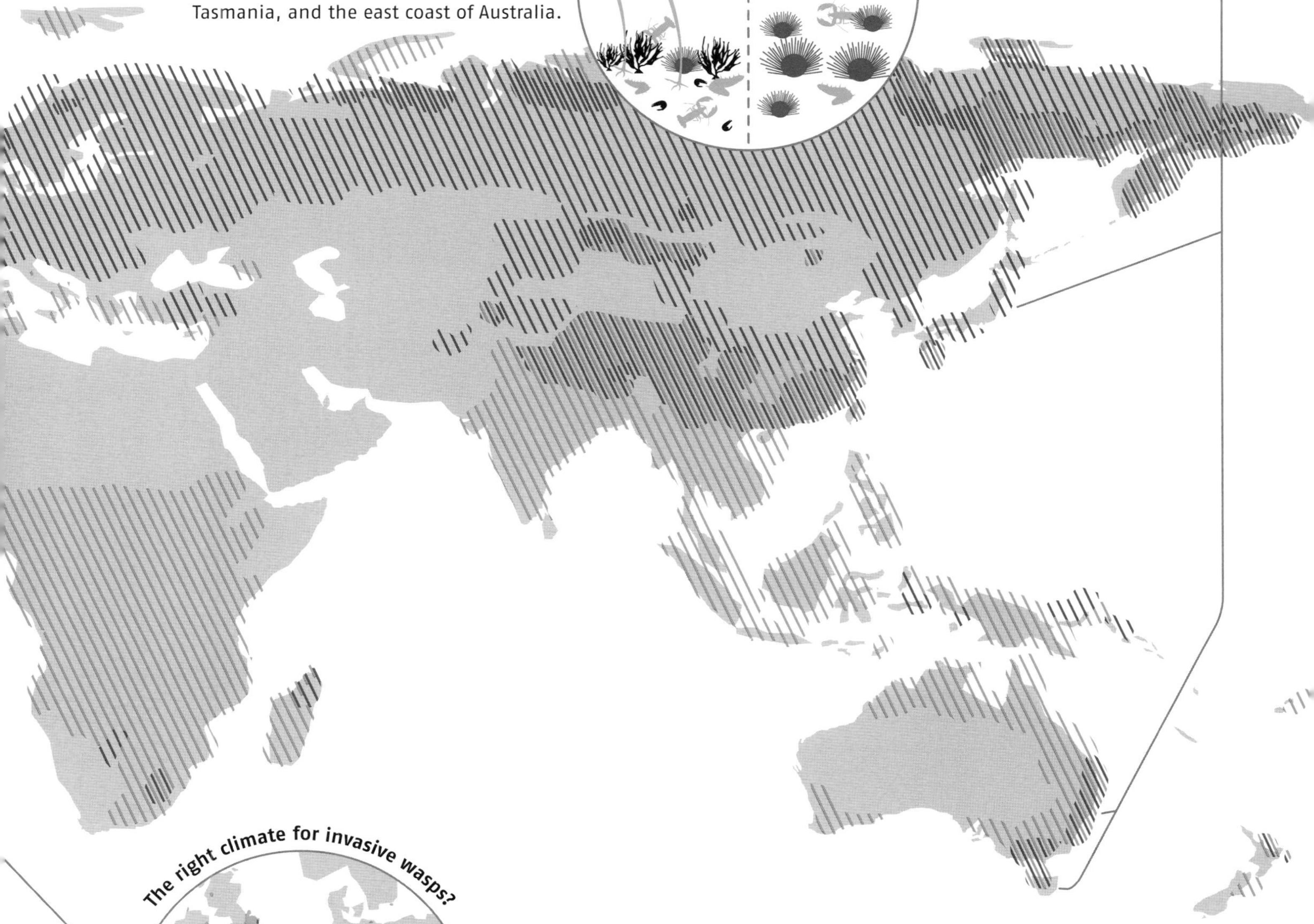

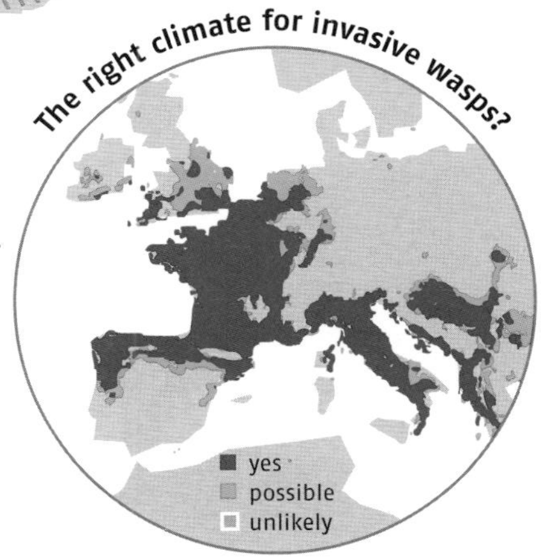

The right climate for invasive wasps?

- yes
- possible
- unlikely

Forecast for 2100

Insects
Asian wasp

Invasive species in France, Italy, and northern Spain are spreading due to the warmer climate. The problem: they are killing native bees.

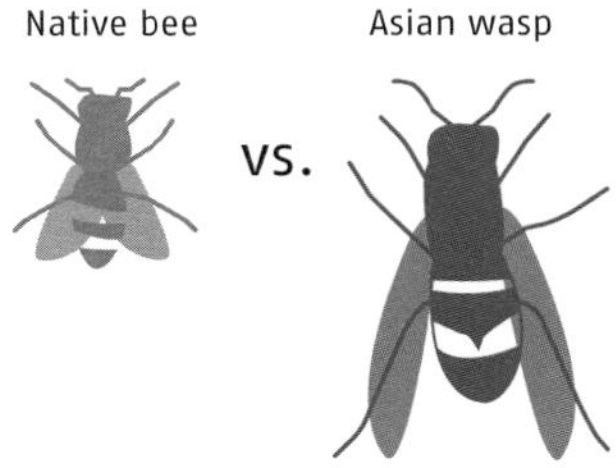

Native bee   Asian wasp

vs.

Sources: Barbet-Massin et al. (2019), Bellard et al. (2013), Ling et al. (2009), Lockwood et al. (2019), Vickery (2020)

Marine Animals

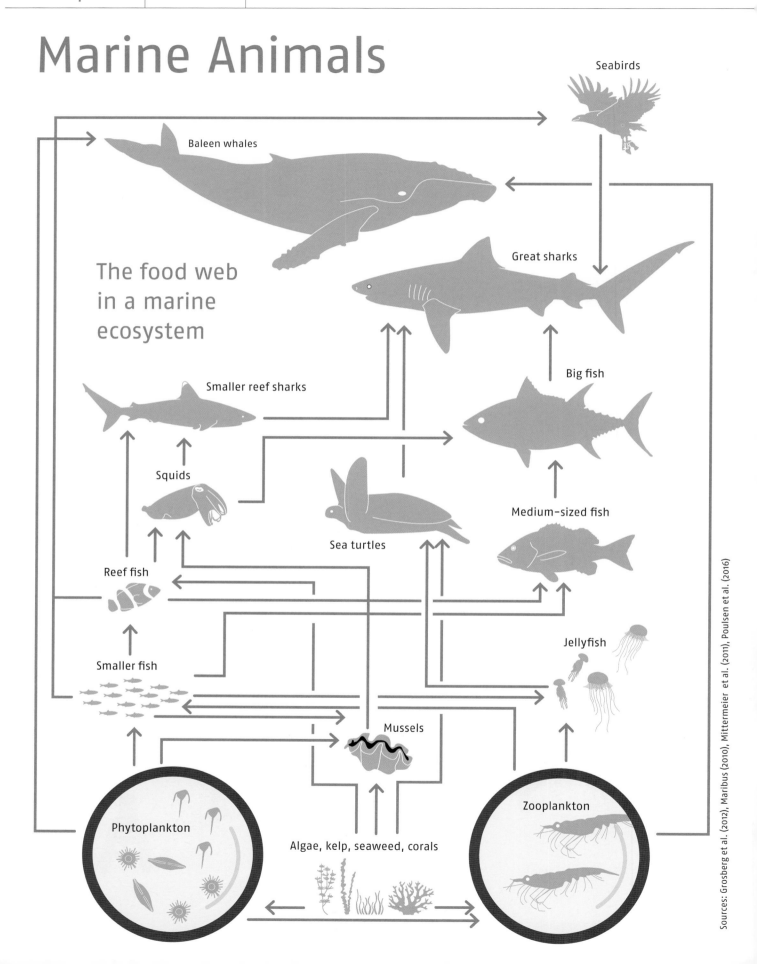

The food web in a marine ecosystem

Seabirds

Baleen whales

Great sharks

Smaller reef sharks

Big fish

Squids

Medium-sized fish

Sea turtles

Reef fish

Jellyfish

Smaller fish

Mussels

Phytoplankton

Zooplankton

Algae, kelp, seaweed, corals

Sources: Grosberg et al. (2012), Maribus (2010), Mittermeier et al. (2011), Poulsen et al. (2016)

Biodiversity comparison land vs. ocean

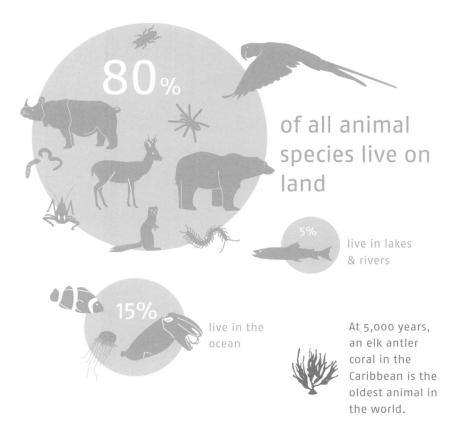

80% of all animal species live on land

5% live in lakes & rivers

15% live in the ocean

At 5,000 years, an elk antler coral in the Caribbean is the oldest animal in the world.

Our oceans are home to countless species of animals and plants. Every day, new members of marine ecosystems are discovered.

Yet as we discover new marine species, we are also losing them. Since industrialization, biodiversity has declined by 65 to 90 percent in some areas.

This is due in part to the destruction of many marine habitats by humans, with bottom trawls plowing up the seabed, coastal wetlands being destroyed by over-fertilization and development, and plastic pollution reaching the remotest and deepest regions of the world's oceans.

The consequences are disastrous: the adaptability to climate change and the productivity of ecosystems decline with every species that goes extinct.

Hotspots of marine biodiversity

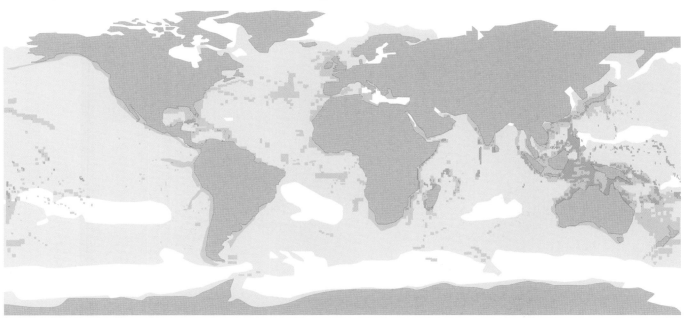

Biodiversity ■ very high (3364–8290*) ■ high (554–3363) ■ medium (92–553) ☐ low (1–91)

*Number of species per 0.5 degrees of longitude and latitude

On the Red List
(due to overfishing and bycatch)

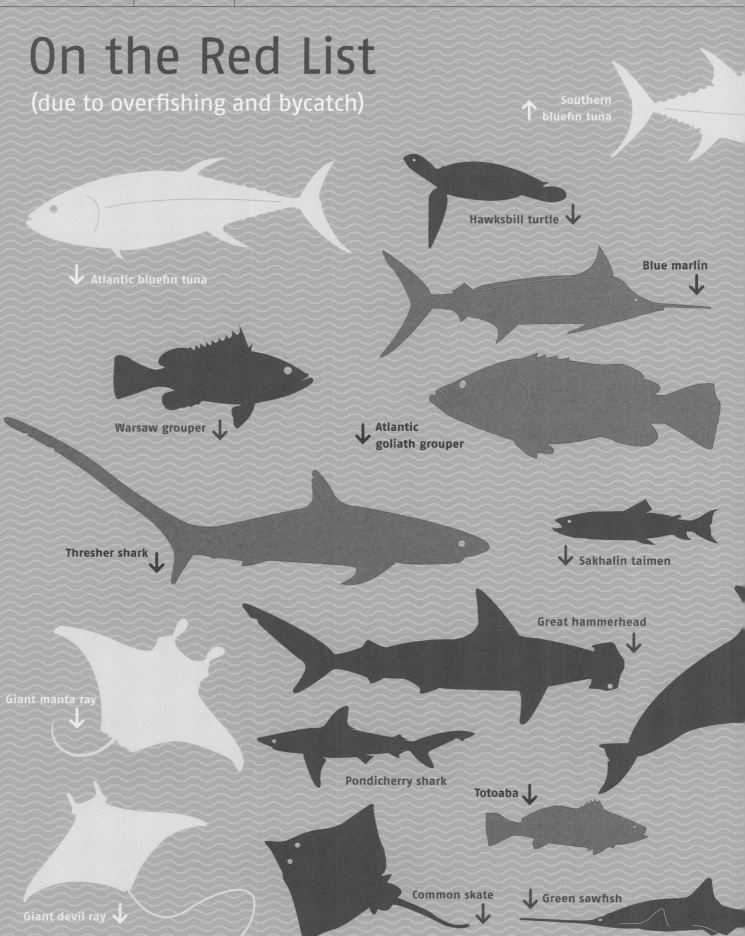

Southern bluefin tuna ↑

Hawksbill turtle ↓

Blue marlin ↓

↓ Atlantic bluefin tuna

Warsaw grouper ↓

↑ Atlantic goliath grouper

Thresher shark ↓

↓ Sakhalin taimen

Great hammerhead ↓

Giant manta ray ↓

Pondicherry shark

Totoaba ↓

Common skate ↓

↓ Green sawfish

Giant devil ray ↓

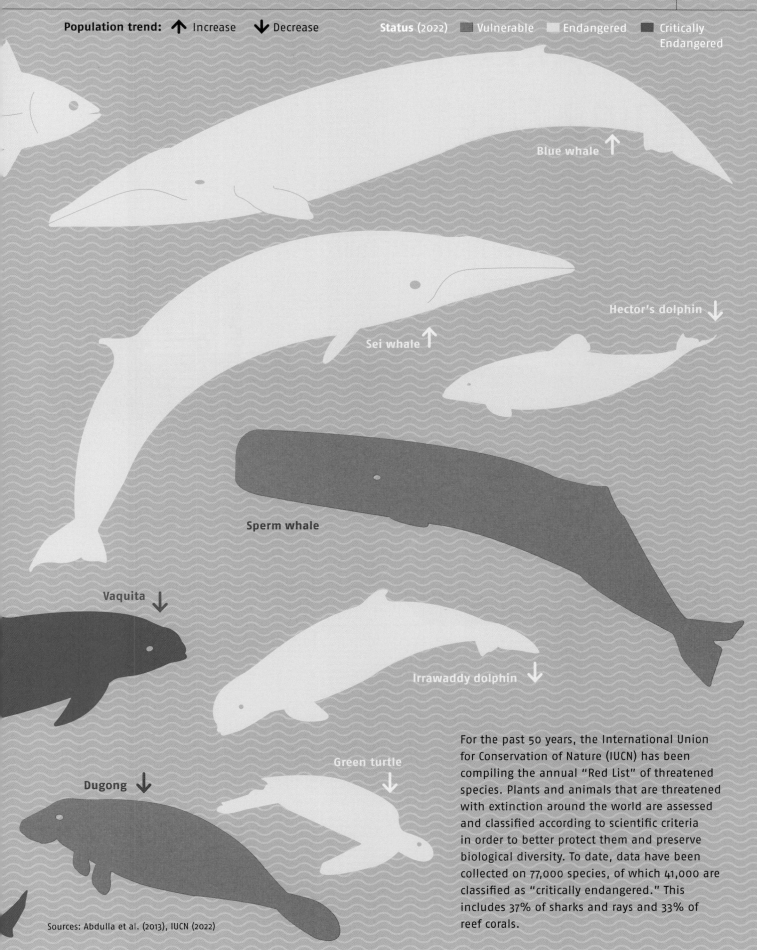

Population trend: ↑ Increase ↓ Decrease   **Status** (2022) ■ Vulnerable ■ Endangered ■ Critically Endangered

Blue whale ↑

Hector's dolphin ↓

Sei whale ↑

Sperm whale

Vaquita ↓

Irrawaddy dolphin ↓

Green turtle ↓

Dugong ↓

For the past 50 years, the International Union for Conservation of Nature (IUCN) has been compiling the annual "Red List" of threatened species. Plants and animals that are threatened with extinction around the world are assessed and classified according to scientific criteria in order to better protect them and preserve biological diversity. To date, data have been collected on 77,000 species, of which 41,000 are classified as "critically endangered." This includes 37% of sharks and rays and 33% of reef corals.

Sources: Abdulla et al. (2013), IUCN (2022)

Oversized Fishing Fleet

The global fishing fleet is estimated at 4.7 million vessels. They have the potential to catch far more fish than fish stocks can sustain.

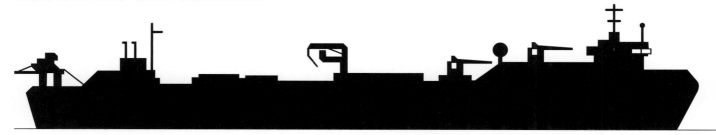

Super trawler with fish factory, 120–144 m,
at sea for several months

Fish trawler with fish factory, 70–90 m,
at sea for approx. 1–2 months

Large fishing vessel, 25–45 m,
at sea for approx. 1–4 weeks

Traditional fishing boat, 7–10 m,
at sea for 1 day

| 0 m | 50 m | 100 m | 140 m |

250,000 people   60,000 kg fish

Maximum catch per trip

7,000,000 kg

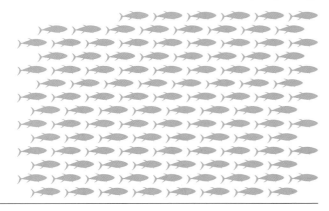

Employment in industrial fishing worldwide

500,000 people

Employment in traditional fisheries

12,000,000 people

1,000,000–1,500,000 kg

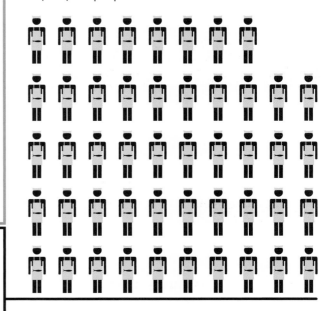

60,000 kg

30–300 kg

Sources: EU (2016), Greenpeace (2014), Reedereien (2017)

Stop Overfishing

Fish stocks are said to be "overfished" when more fish are caught on a sustained basis than can be replenished through migration or natural reproduction. This can be seen, for example, when catches continue to fall despite improvements in fishing techniques, longer fishing trips, and increased fishing effort.

Some 30–55% of stocks are overfished or collapsed, according to scientific estimates and calculations.

A stock is "collapsed" when it declines disproportionately within a few years and has little chance of recovering. To achieve sustainable fishing, scientifically-based catch quotas must be introduced and enforced worldwide.

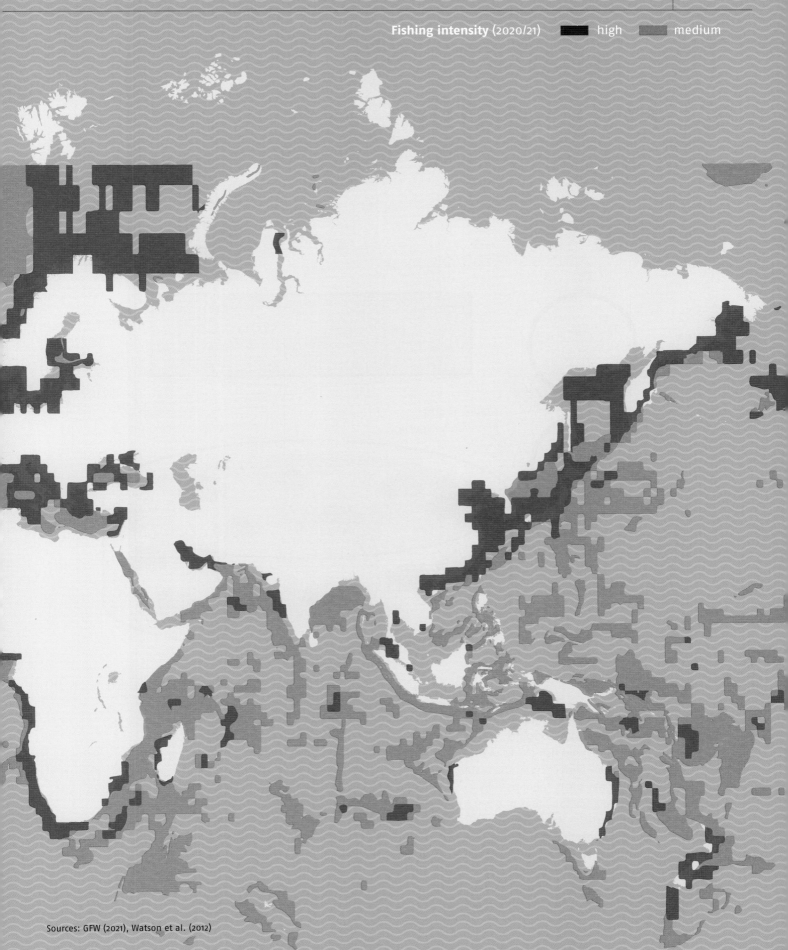

Fishing intensity (2020/21) ■ high ■ medium

Sources: GFW (2021), Watson et al. (2012)

The Future of Fish Farming

Aquaponics

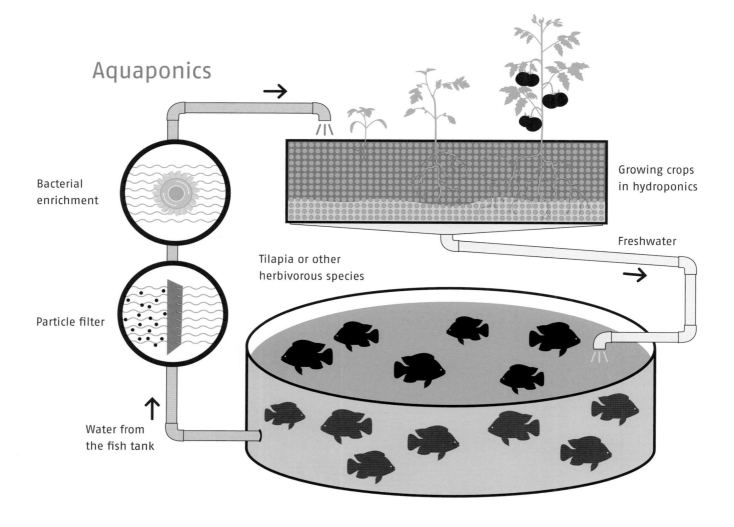

Bacterial enrichment

Particle filter

Water from the fish tank

Tilapia or other herbivorous species

Growing crops in hydroponics

Freshwater

While not all fish farming is sustainable, some approaches follow closed-loop concepts, reducing the pressure on wild fish stocks. Aquaponics can create a near-perfect cycle: the water from the tank, which has been "polluted" by the fish, is pumped to grow tomatoes or other plants. The coarsest particles are removed and bacteria are added to make it easier for the plants to absorb nutrients. The resulting natural fertilizer feeds the plants, making them grow faster and more productively, with the positive side effect of purifying the water, which is fed directly back to the fish.

The plants add enough nutrients to the water for the fish to thrive, which means that in a well-organized farm, fish feeding can be drastically reduced or even eliminated. The use of plants also lowers CO_2 emissions by reducing the need for power-intensive water filtration technology. Fewer fish per tank and higher standards of hygiene also reduce stress and disease, eliminating the need for prophylactic antibiotics.

Sources: DFO (2013), Maribus (2013)

Integrated Multi-Tropic Aquaculture (IMTA)

In this system, caged fish are the only species which are fed directly. Their excretions and leftovers provide nutrients for a whole slew of other organisms.

Fish excrement and food leftovers float over to bivalves and seaweed grown next to the fish enclosure, where nutrients are absorbed in turn, with food particles filtered out from the "waste."

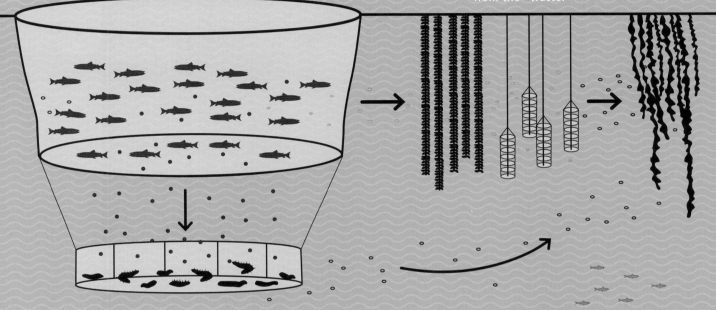

Some of the excess fish food and excrement sink down and are eaten by sea cucumbers held in a cage below the fish.

IMTA alleviates negative side effects associated with aquaculture, such as over-fertilization, keeping the local ecosystem around farms in balance. This new type of aquaculture is well on its way to shaping the future of the industry, because it is also more economical: equal amounts of feed and space support higher yields and allow a wider variety of fish to be raised and sold.

The system follows the principles of permaculture the opposite of monoculture, meaning that instead of just farming a single species, an entire community is raised. Researchers are now trying to identify which additional organisms can complete an IMTA system.

The last of the nutrients sink to the ocean floor, where they can increase algae growth. This algae coverage is then grazed on by sea urchins or other organisms.

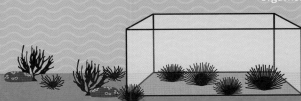

Noisy Oceans

Noise levels in the oceans have doubled every decade since 1950 due to increased shipping and seismic exploration, which produces sounds as loud as a nuclear bomb and is used to find fossil fuels.

In recent years, shipping activity in Canada's eastern Saint Lawrence River has increased, and the beluga whale population has begun to decline. To counteract this, part of the river is now a protected area.

Fishing fleet sonar

2020

2010

2000

1990

1980

1970

1960

1950

Orcas around Vancouver Island are so stressed by ship traffic that they increase their swimming speed and leave the area when more than one ship is nearby. This causes them to use up valuable energy that they could otherwise use to find food.

km/h

Sources: Duarte (2021), Nowacek et al. (2001), Schorr et al. (2014), Veirs et al. (2016)

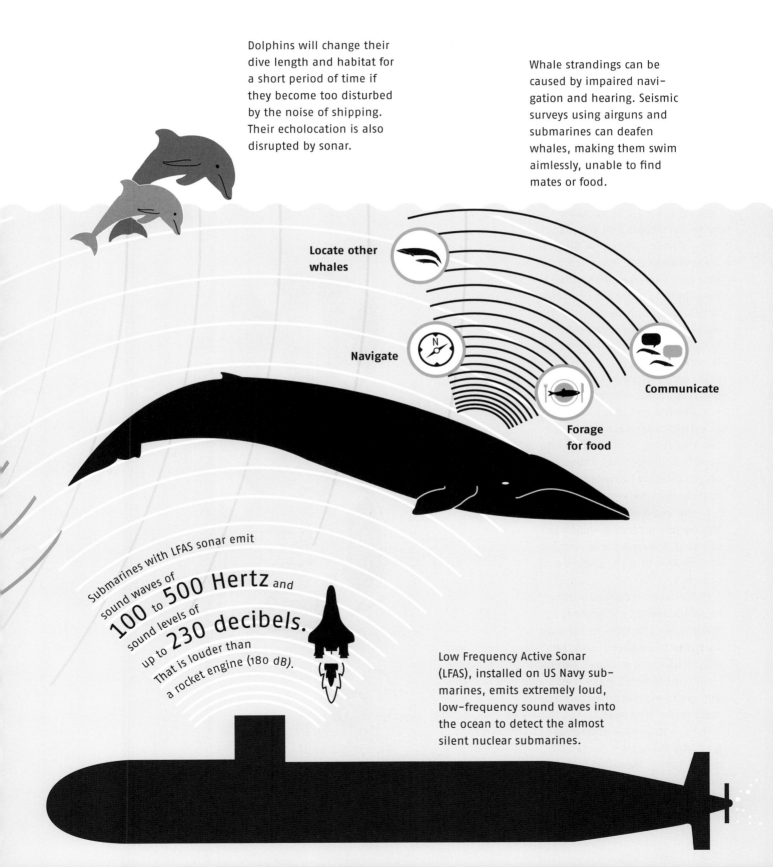

Dolphins will change their dive length and habitat for a short period of time if they become too disturbed by the noise of shipping. Their echolocation is also disrupted by sonar.

Whale strandings can be caused by impaired navigation and hearing. Seismic surveys using airguns and submarines can deafen whales, making them swim aimlessly, unable to find mates or food.

Locate other whales

Navigate

Communicate

Forage for food

Submarines with LFAS sonar emit sound waves of **100 to 500 Hertz** and sound levels of up to **230 decibels.** That is louder than a rocket engine (180 dB).

Low Frequency Active Sonar (LFAS), installed on US Navy submarines, emits extremely loud, low-frequency sound waves into the ocean to detect the almost silent nuclear submarines.

Marine Protected Areas Are Growing

To ensure a more sustainable future for people and our planet, the expansion of marine protected areas is essential. Protected areas conserve biodiversity and thus contribute to a healthy ecosystem, which, in turn, is better able to withstand human intervention, climate change, and the problems it causes.

The most effective are large protected areas where fishing is not allowed and that have been in existence for a long time. In the long term, such reserves also benefit fishermen, who can count on consistently higher catch quotas outside the reserves.

In the last 6 years, the size of the world's protected areas has doubled. They now cover some 32 million km², or 8.1% of the world's oceans.

While protected areas in national territorial waters are growing steadily, growth rates in the high seas, where no government has jurisdiction, are minimal. Complex legal requirements make it very difficult to establish new protected areas in the high seas.

The top 10 marine protected areas are Palau (78% protected), the UK (39%), Mauritius (29%) and the USA (24%). They are followed by up to 12%: Chile, Kiribati, Argentina, Australia, Mexico and Ecuador.

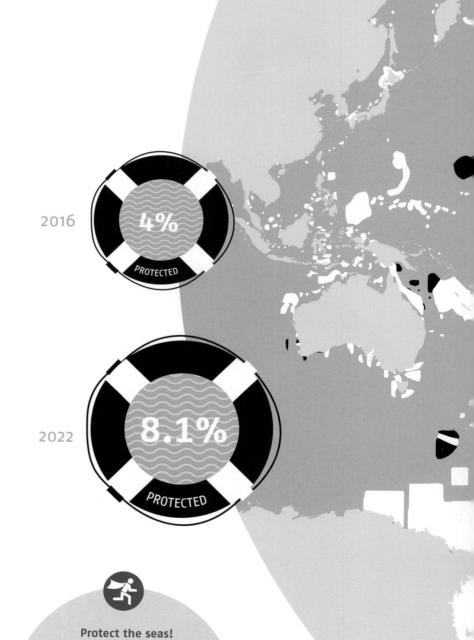

2016 — 4% PROTECTED

2022 — 8.1% PROTECTED

Protect the seas!
The protected areas are growing steadily, and regularly Updated information on protected areas is available here:
mpatlas.org

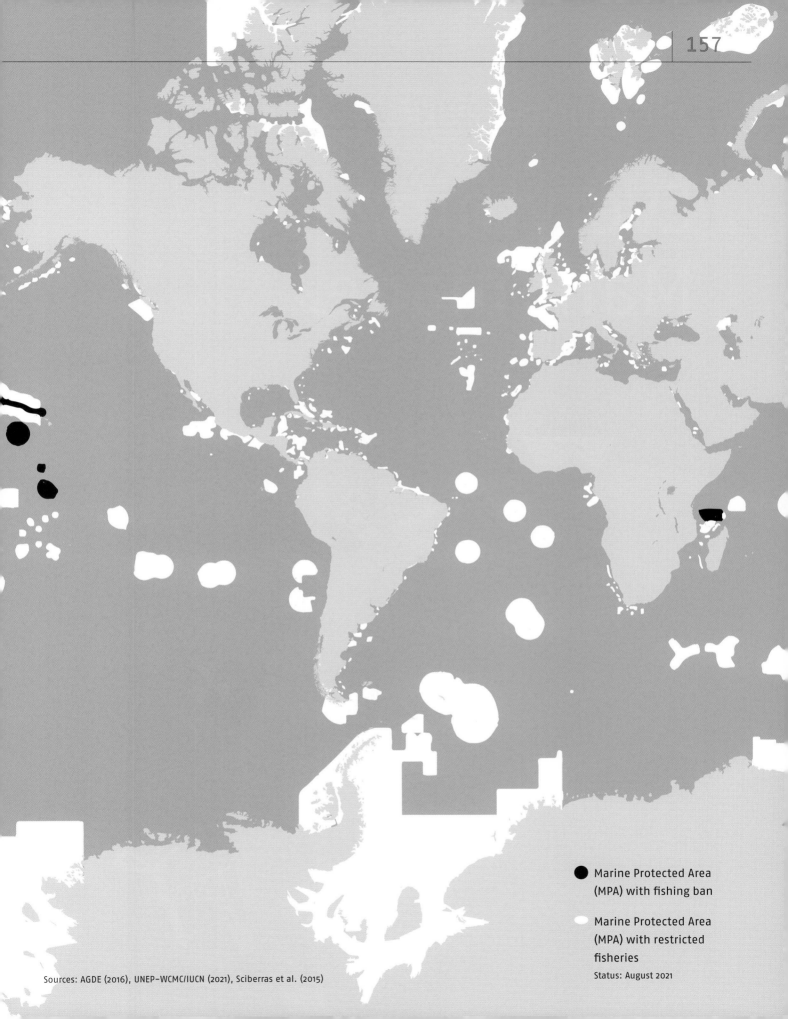

Marine Protected Area
(MPA) with fishing ban

Marine Protected Area
(MPA) with restricted
fisheries

Status: August 2021

Sources: AGDE (2016), UNEP–WCMC/IUCN (2021), Sciberras et al. (2015)

Everything
Man-made

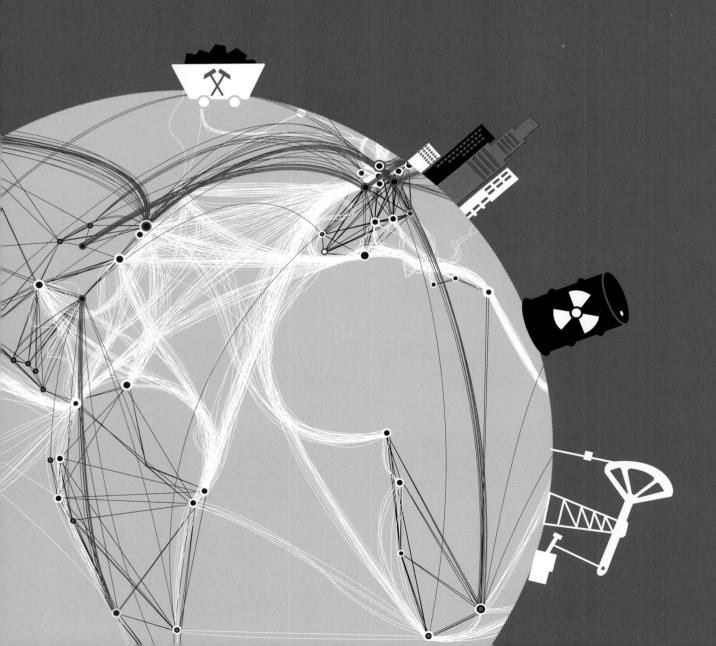

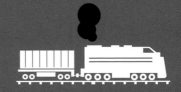

An | thro | po | sphere
[Ancient Greek anthropos = human
& sphaîra = sphere (Earth)]

The anthroposphere includes everything
that has been created, modified, or
degraded by humans. This includes
changes to the Earth's surface for
agriculture, transport, industry, or
buildings, as well as chemical changes to
the air, water, and soil.

Raw Materials & Industry

Global Raw Material Extraction

Raw material processing

Worldwide

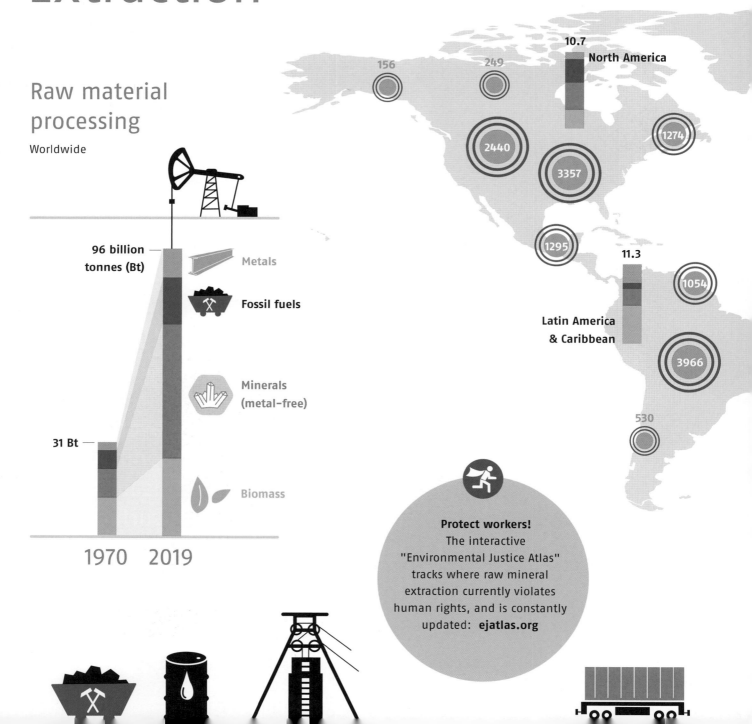

96 billion tonnes (Bt)

Metals

Fossil fuels

Minerals (metal-free)

31 Bt

Biomass

1970   2019

156

249

10.7   North America

2440

3357

1274

1295

11.3

1054

Latin America & Caribbean

3966

530

Protect workers!
The interactive "Environmental Justice Atlas" tracks where raw mineral extraction currently violates human rights, and is constantly updated: **ejatlas.org**

Where raw materials are mined

50.6 Bt

Asia & Pacific

Number of mines (global hotspots)

8.7 — Europe

5.8 — Eastern Europe & Central Asia

7.8 — Africa

2.9 — East Asia

319

891

798

1786

2085

2722

1119

60

736

3569

150

6209

1342

3298

2122

2915

370

The extraction and processing of raw materials is responsible for around half of the world's CO_2 emissions.

It is a huge factor that is not given enough attention when it comes to reducing emissions. According to the United Nations Environment Programme (UNEP), resource extraction is set to double by 2050.

The urgently needed reduction in the use of raw materials can be achieved through innovative and efficient processing technology and by moving towards a circular economy (see p. 180).

If raw materials are seen as limited resources that can be recycled and reused after their initial use, large amounts of CO_2 can be saved.

Source: WU Vienna (2022)

Deep Sea Mining

Many raw materials of interest to industry are found in the seabed. These include oil and gas, but also manganese nodules, massive sulfides, and cobalt crusts.

Manganese nodules form on the seabed at depths of 4000 to 6500 meters and take millions of years to grow to just a few millimeters. The 5 to 10 cm nodules contain valuable industrial metals such as manganese, iron, nickel, copper, lithium, and cobalt.

Massive sulfides form where continental plates drift apart, at depths of 500 to 5000 meters. They contain copper, zinc, and gold.

So far, only a few places are known where mining would be economically viable, such as the Manus Basin in the southwest Pacific.

Occurrence below 400 m sea depth ● Crude oil ● Natural gas ▲▲ Massive sulfides ▨ Manganese nodules ∥∥ Cobalt crusts

Cobalt crusts grow even more slowly than manganese nodules, at all water depths, on sediment-free rock surfaces such as rocky slopes or seamounts. They consist mainly of manganese and iron and contain small amounts of valuable elements such as nickel, cobalt, copper, titanium, and rare earths. The latter are used in mobile phones, flat screens, hybrid vehicles, and wind turbines.

Deep-sea mining of manganese nodules, cobalt crusts, and massive sulfides is still theoretical. Although there are prototypes of mining machines that can operate at depths of 2000 meters, long-term exploitation at depths of up to 6000 meters remains technically unsolvable. This is a good thing, because the damage to the seabed is visible to the naked eye. As a team of researchers recently discovered, the seafloor from the first manganese nodule mining trials in the Pacific is still visibly scarred after 25 years.

The extraction of oil and gas from the sea accounts for about a quarter of the world's production. Offshore drilling takes place mainly in the Middle East, the North Sea, off Brazil, the Gulf of Mexico, and the Caspian Sea.

Sources: GEOMAR (2016), IEA (2018), Janssen et al. (2020), Maribus (2014), Rona (2003), UBA (2013), UNEP (2013)

The Biggest Oil Spills 1901–2020

Persian Gulf
During the Gulf War in January 1991, Iraqi soldiers opened the valves of the Sea Island oil terminal while the US military fired on Iraqi oil tankers. Around 1,000,000 tonnes of oil spilled, contaminating the coasts of southern Kuwait and Saudi Arabia.

USA
In March 1909, the largest oil spill in history occurred near Kern County, California. Approximately 1,227,600 tonnes of oil were released into the environment from a drilling leak.

USA
In April 2010, the deep-sea platform "Deepwater Horizon" spilled around 470,779 tonnes of heavy oil in a single leak.

Sources: CEDRE (2016), Maribus (2010), safety4sea (2020), NA (2022), WHOI (2011)

1 liter of oil can contaminate up to 1 million
liters of drinking water. Worldwide, an estimated
1400 billion tonnes of oil enter the oceans
each year.

How does oil end up in the oceans?

Estimates in million tonnes per year
(2010–2019)

Japan
In November 1974,
52,836 tonnes of oil were
washed into the sea in
Tokyo Bay, off the island
of Honshu, as a result of a
tanker collision.

1,200,000 Mt

Wastewater containing oil,
for example from roads and
industrial plants, reaches
the oceans via rivers and
sewers.

Natural sources,
like oil seeping from
the seabed.

100,000

Oil production,
such as oil loss during
normal operation and in
the event of accidents.

66,500

South Africa
In June 1983, a fire on the
Spanish tanker "Castillo
de Bellver" off the coast of
Saldanha Bay caused a spill
of some 250,000 tonnes of
light oil.

Transport of oil,
such as tanker accidents.
 818

Other oil spills 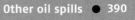 390

Raw Materials for Energy Needs

As one of the world's leading industrial nations, the US consumes vast quantities of raw materials. Mineral raw materials such as stone and earth are largely extracted from US open-cast mines and quarries. As far as energy raw materials are concerned, the US is partly dependent on imports.

The Russian invasion of Ukraine shows that our dependence on imports must be reduced and the diversity of supplier countries increased.

For the sake of energy independence and climate protection, the focus should also be on expanding renewable energies about seven times faster than currently.

Electricity generation in the US

2023

Renewable energies
21.4%

Coal
16.2%

Natural gas
43.1%

Other
0.7%

Nuclear power
18.6%

Imported fuels
US, 2021

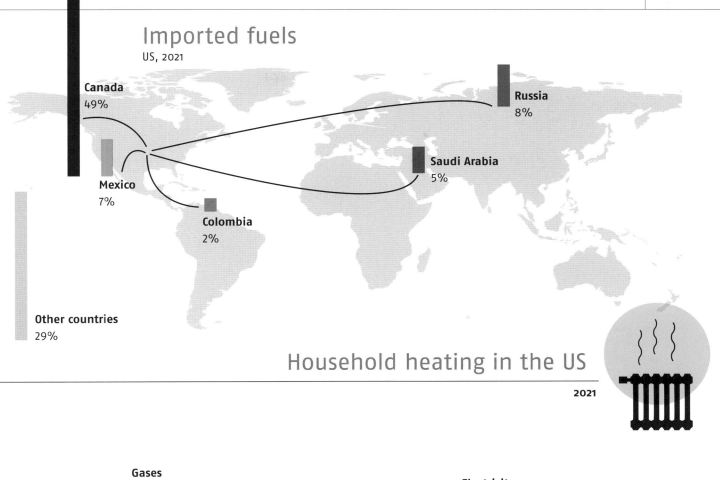

Canada
49%

Russia
8%

Mexico
7%

Saudi Arabia
5%

Colombia
2%

Other countries
29%

Household heating in the US
2021

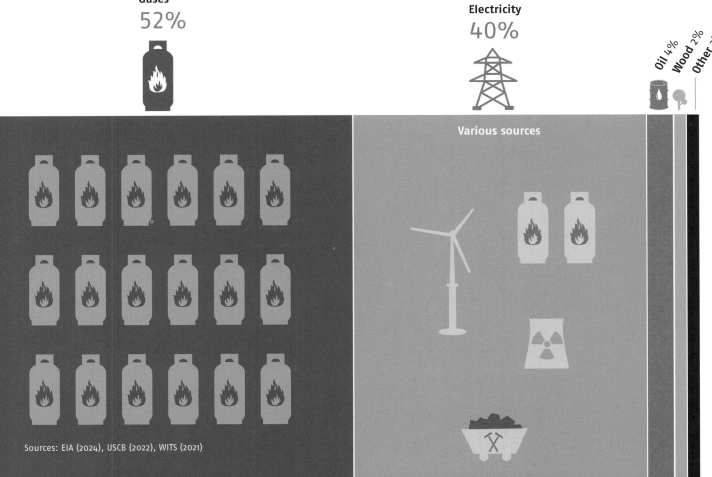

Gases
52%

Electricity
40%

Oil 4% Wood 2% Other 2%

Various sources

Sources: EIA (2024), USCB (2022), WITS (2021)

Radioactive Raw Materials

A high-risk technology on the descent

10%

Canada ranks third in uranium production, with the world's largest uranium mine at Cigar Lake. About 10% of the world's demand is mined underground there: 4693 tonnes.

Uranium mining
2021, worldwide

55

nuclear power plants are connected to the grid in the US (2021), producing 19% of US electricity demand and falling.

Germany has phased out nuclear power, closing its last three plants in 2023.

All three power plants had been in operation since 1988.

70%

nuclear power in the electricity mix: France has the highest nuclear dependency in the world.

45%
of the world's uranium
comes from Kazakhstan:
21,819 tonnes per year.

Finland has
created the
world's first
permanent
storage facil-
ity* for spent
nuclear fuel.

* A cave 450 m underground
in the bedrock. Unlike
above-ground sites it is safe
from bushfires, flooding,
and terrorist attacks.

Uzbekistan is in 5th
place. It mines 3500
tonnes of uranium.

12%
Namibia is the second
largest producer of
uranium in the world
with 5753 tonnes.

9%
Australia is in 4th place
with 4192 tonnes of
uranium mined.

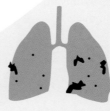

Uranium miners
have a 6 times
higher rate of lung
cancer* and a 24
times higher risk of
other lung diseases.

*Compared to other miners

Sources: BMUV (2022), CDC (2020), EIA (2022), WNA (2022)

Trees as "Green" Energy Sources?

Vital for survival
65 million refugees worldwide depend on wood as their only accessible source of energy.

40%

40% of global renewable energy currently comes from wood. Much of this production is not sustainable or climate-friendly, however. The long-term benefits of energy from wood products depend on local, sustainable forestry practices and new technology, and aren't on their own a climate "solution."

Hydropower helpers
Forested tributaries ensure a longer service life for hydroelectric turbines by naturally filtering the water and mitigating the negative effects of droughts.

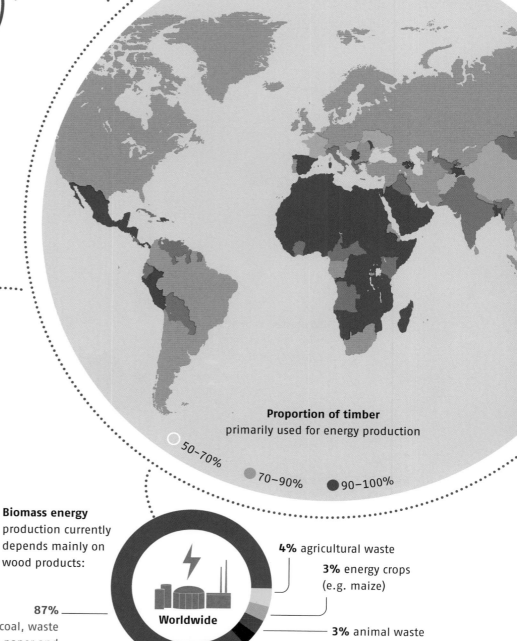

Proportion of timber primarily used for energy production

○ 50–70%   ● 70–90%   ● 90–100%

Biomass energy production currently depends mainly on wood products:

4% agricultural waste

3% energy crops (e.g. maize)

87% wood, charcoal, waste from the paper and timber industry, etc.

Worldwide

3% animal waste

3% waste from landfills

Transformation of conventional forests into sustainable, near-natural commercial forests

Increase political incentives for long-term investments in bioenergy

2.4 billion people, about a third of the world's population, use wood every day for cooking, heating, or sterilizing drinking water. People in Africa are particularly dependent on wood.

Greater international cooperation in the regulation of the timber trade and the transfer of technology and knowledge

50% of the world's round timber, 1.86 billion cubic meters, is used for energy purposes.

Expanding help: developing cleaner, cheaper, and more efficient ovens

Establish training programs

Promoting sustainable forest management worldwide

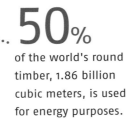 

Most wood pellets are pressed from waste produced in sawmills. However, this is not always the case: in the USA, wood pellet exports have increased since 2013 due to the subsidization of wood pellet power plants in the EU. Entire forests are being cut down for this purpose. In this case, pellets have a worse carbon footprint than coal.

Even if the forest cut down for pellets is immediately reforested, it will take 40–100 years for the newly grown forest to have a comparable positive effect on the climate – far too late for the acute climate crisis!

US export of wood pellets
in million tonnes per year

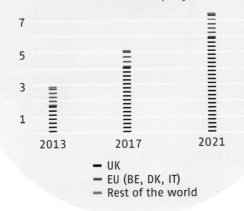

7
5
3
1
2013   2017   2021

— UK
— EU (BE, DK, IT)
— Rest of the world

Long transport routes and deforestation: is this form of energy production really "carbon neutral"?

Sources: FAO (2018 & 2020), Goncalves et al. (2018), Sterman et al. (2018), USITC (2018), Voegele (2022), WBA (2017)

The Future of Energy is Green!

The energy supply of the future
– electricity, heat, and transport – will be based on thousands of small to gigantic photovoltaic and wind power plants.

Hydropower, hydrogen, biomass, and geothermal energy will also play a role in the energy mix. Energy producers will be "dispersed" across a country, meaning that electricity supply will become more decentralized, complemented by many regional electricity storage facilities.

Wherever possible, **industrial plants** organize their processes flexibly to use electricity when it can be produced.

Regional electricity storage
(such as pumped storage plants or battery energy storage systems) feed the energy back into the grid when required.

Biomass

Transportation revolution
An important goal alongside electrification is the reduction of private transportation. Strategies include expanding local and long-distance transportation and improving the infrastructure for bicycles and pedestrians.

Transportation transition

E-car sharing
Car sharing is important for anyone who can't take the bus or train. The fuller the car, the better.

Ride more
There is a growing range of sports bikes, cargo bikes, folding bikes and trailers – even electric ones if you have a lot of luggage or have long journeys.

100% Renewable Energy by 2050 (Projected energy mix for 139 countries)

57.5% Solar

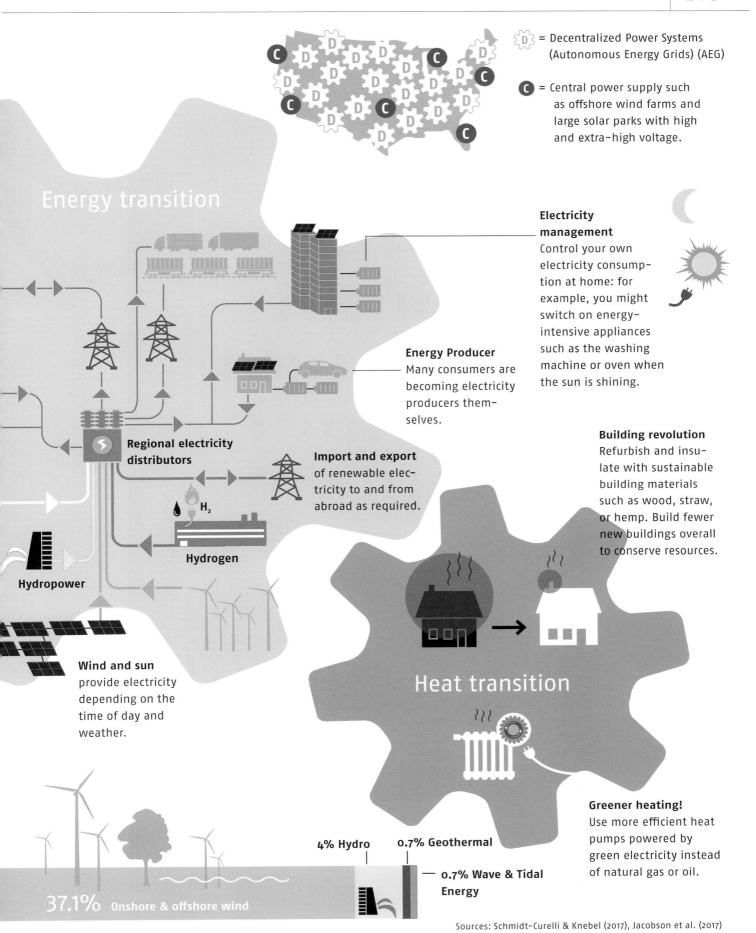

= Decentralized Power Systems (Autonomous Energy Grids) (AEG)

= Central power supply such as offshore wind farms and large solar parks with high and extra-high voltage.

Energy transition

Electricity management
Control your own electricity consumption at home: for example, you might switch on energy-intensive appliances such as the washing machine or oven when the sun is shining.

Energy Producer
Many consumers are becoming electricity producers themselves.

Building revolution
Refurbish and insulate with sustainable building materials such as wood, straw, or hemp. Build fewer new buildings overall to conserve resources.

Regional electricity distributors

Import and export
of renewable electricity to and from abroad as required.

H_2

Hydrogen

Hydropower

Heat transition

Wind and sun
provide electricity depending on the time of day and weather.

Greener heating!
Use more efficient heat pumps powered by green electricity instead of natural gas or oil.

4% Hydro **0.7% Geothermal**

0.7% Wave & Tidal Energy

37.1% Onshore & offshore wind

Sources: Schmidt-Curelli & Knebel (2017), Jacobson et al. (2017)

Sustainable Forestry

In the US, forests grow more wood than is removed or lost to mortality. Annual net growth (i.e., growth minus mortality) is about twice annual harvests.

1/3 of the land in the United States is classified as forest, constituting the fourth largest forest land base of any nation worldwide.

950,000 employees work in the U.S. forest products industry.

Sources: McGinley (2023), USFS (2022)

56% of US forest lands are in private ownership.

44% are managed by local, tribal, state, and federal governments.

70% of US timberland   30% of timberland

Wood Production in Transition

Worldwide

21st century
Modern use of wood
Bioenergy (wood pellets),
biochemistry (nanocellulose),
textiles, car parts, building
materials, and much more.

20th century
Traditional use of wood
Paper/pulp, building
materials, wood
products like furniture,
and heating.

27 Mt of wood were recycled globally in 2018. One of the UN's sustainability goals is to further expand the circular economy in the wood sector by 2030.

Sources: FAO (2019 + 2020), Hetemäki & Hurmekoski (2016), IEA (2022), UNECE (2019)

Paper production
1990–2021

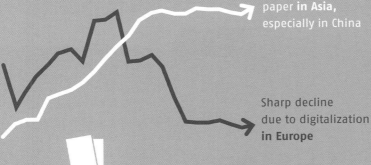

Rising demand for paper **in Asia,** especially in China

Sharp decline due to digitalization **in Europe**

Record year 2021
Global paper production reached a new record of 415 million tonnes in 2021, after production leveled off slightly in 2018–2020 due to the pandemic and a decline in newspaper and magazine printing. Population growth will tend to increase paper consumption. Over the past two decades, China has become the world's largest producer and consumer of paper and pulp.

Built out of Sand

Sand is made up of 0.063–2 mm small fragments that have been loosened from substances like rock, coral, or shells over thousands or millions of years by abrasion and weathering. It is carried by the wind as dust over long distances, forming large deserts such as the Sahara.

Sand is made every day in the mountains, in rivers, or in the ocean. However, we humans use far more sand than nature "produces" through weathering.

Our civilization is literally built on sand: the only thing we use more of is water. Every person on Earth uses an average of 15 kg per day. If you were to build a wall with the amount of sand that we use in a year, it would be 27 meters high, 27 meters wide, and would go around the world once.

A surprising number of products require sand. Among other things, it is used for...

An estimated 50 bn tonnes of sand are mined worldwide every year.

Sources: Dewi et al. (2019), Langrand & Peduzzi (2022), Masalu (2010), UNEP (2022)

Construction of buildings, e.g. concrete, brick, plaster

Infrastructure, e.g. roads and bridges

Energy production, e.g. solar panels

Technology & Internet, e.g. computer chips, fiber optic cable

Electricity storage, e.g. lithium batteries

Glass products, e.g. glass, windows, mirrors

Hygiene products, e.g. toothpaste

Textiles, e.g. jeans

Sand mining

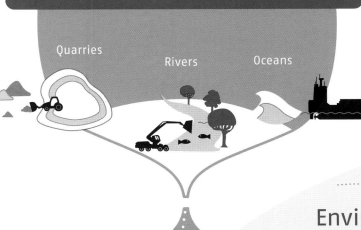

Quarries · Rivers · Oceans

Approx. **15 %** of the world's sand is mined illegally. In Morocco, half of the beaches have been illegally removed; in Tanzania, illegal mining is causing coastal erosion; in India, there is even a sand mafia.

Environmental impact

Destruction of marine habitats and ecosystems. The removal of micro-organisms and plants that live in the sand, such as seagrass, means there is less food and breeding grounds for fish, and biodiversity is drastically reduced.

Rivers & Deltas
Sand mining causes river banks to collapse, destroys breeding grounds, leads to flooding, depletes fish stocks, and can exacerbate drought and poverty.

Sustainable use

In 2019, the United Nations Environment Programme (UNEP) and member nations recommended steps to make sand use more sustainable worldwide.

Collect more data on sand mining.

Transfer knowledge and improve international networking.

Create regulations for sand utilization and introduce a global standard.

Reduce extraction, especially from dynamic ecosystems such as rivers, beaches, and oceans.

Optimize concrete use and recycling.

Replace sand, using innovative materials such as iron mining residues.

A World of Plastic

403 Mt

of plastic was produced world-wide in 2022. The mountains of waste are growing every day: much of the plastic produced since 1950 can still be found in its original form in landfills, the countryside, rivers, and the world's oceans.

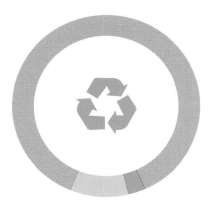

3–10%

of the plastic collected worldwide is recycled.

9 million
barrels of crude oil are used daily in plastic production.

oil
1 day

Plastic that is thrown away after a single use: **33%**

Estimated decomposition time

Paper bags
6 weeks

Plastic bags
1–20 years

Foamed plastic
packaging
50 years

Aluminium cans
200 years

PET bottles
200–450 years

Six-pack rings
400 years

Fishing line
600 years

It takes approx. **450 years** for a PET bottle to break down in the ocean into smaller and smaller pieces until they are invisible to the naked eye. Plastic is not biodegradable and will never disappear completely.

More than
100,000
marine mammals and millions of seabirds and fish die every year from pieces of plastic that have been eaten or wrapped around them.

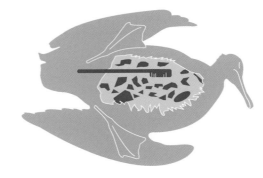

Sources: Schlining et al. (2013), Petroleum Economist (2019), Plastics Europe (2016), Subba Reddy (2014), UNEP (2015), WEF (2016)

The Future of Raw Materials Is Circular!

The world's resources are finite. Many important raw materials will become scarce in the future, while population and demand continue to grow.

In the EU, each citizen consumes an average of 14 tonnes of raw materials per year and produces 5 tonnes of waste. In the current "linear economy," most of this waste is disposed of in landfills and only a small proportion (around 8%) is recycled.

In a modern circular economy, in contrast, almost everything is recycled, reused or repaired.

The life cycle of products is extended, for example through intelligent design with replaceable and repairable elements. At the end of their life, manufacturers are required to take back and recycle their products. Disposable products and low-cost production, such as plastic products, textiles, or electronics, will be minimized.

In a circular economy, fewer raw materials are extracted overall. This reduces CO_2 emissions and the destruction of nature.

Material cycles like this mimic nature's cradle-to-cradle system, where there is no such thing as worthless "waste": everything has value and is used in many different ways.

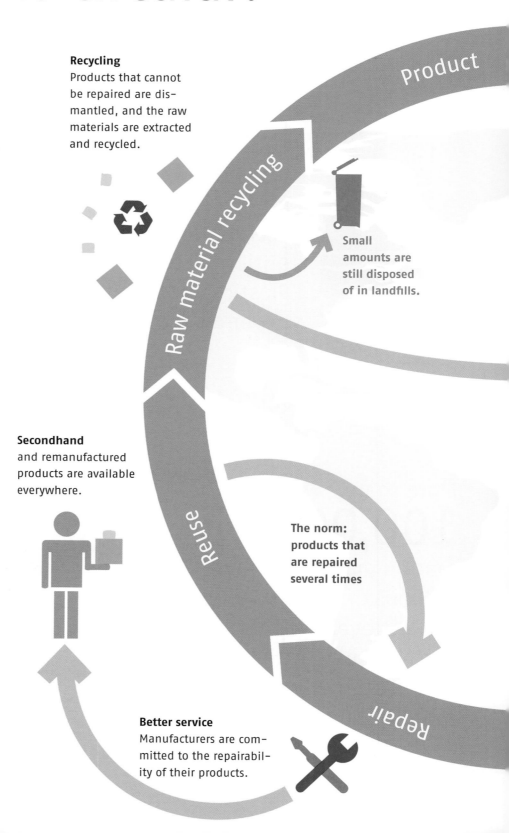

Recycling
Products that cannot be repaired are dismantled, and the raw materials are extracted and recycled.

Small amounts are still disposed of in landfills.

Secondhand
and remanufactured products are available everywhere.

The norm: products that are repaired several times

Better service
Manufacturers are committed to the repairability of their products.

Product

Raw material recycling

Reuse

Repair

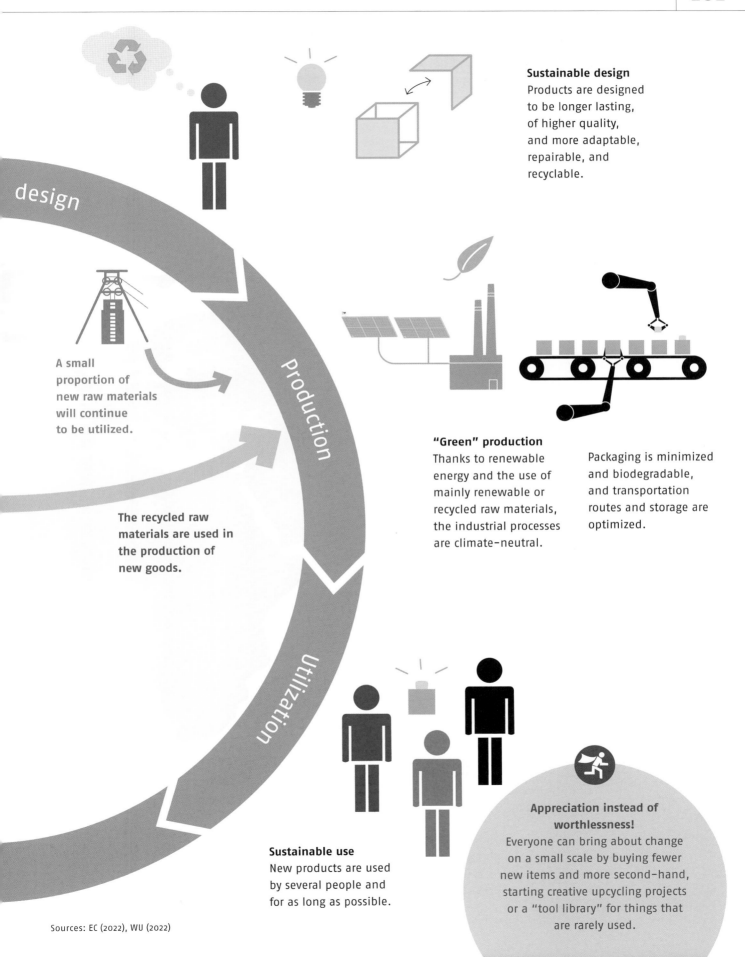

design

Production

Utilization

Sustainable design
Products are designed
to be longer lasting,
of higher quality,
and more adaptable,
repairable, and
recyclable.

A small
proportion of
new raw materials
will continue
to be utilized.

The recycled raw
materials are used in
the production of
new goods.

"Green" production
Thanks to renewable
energy and the use of
mainly renewable or
recycled raw materials,
the industrial processes
are climate-neutral.

Packaging is minimized
and biodegradable,
and transportation
routes and storage are
optimized.

Sustainable use
New products are used
by several people and
for as long as possible.

**Appreciation instead of
worthlessness!**
Everyone can bring about change
on a small scale by buying fewer
new items and more second-hand,
starting creative upcycling projects
or a "tool library" for things that
are rarely used.

Sources: EC (2022), WU (2022)

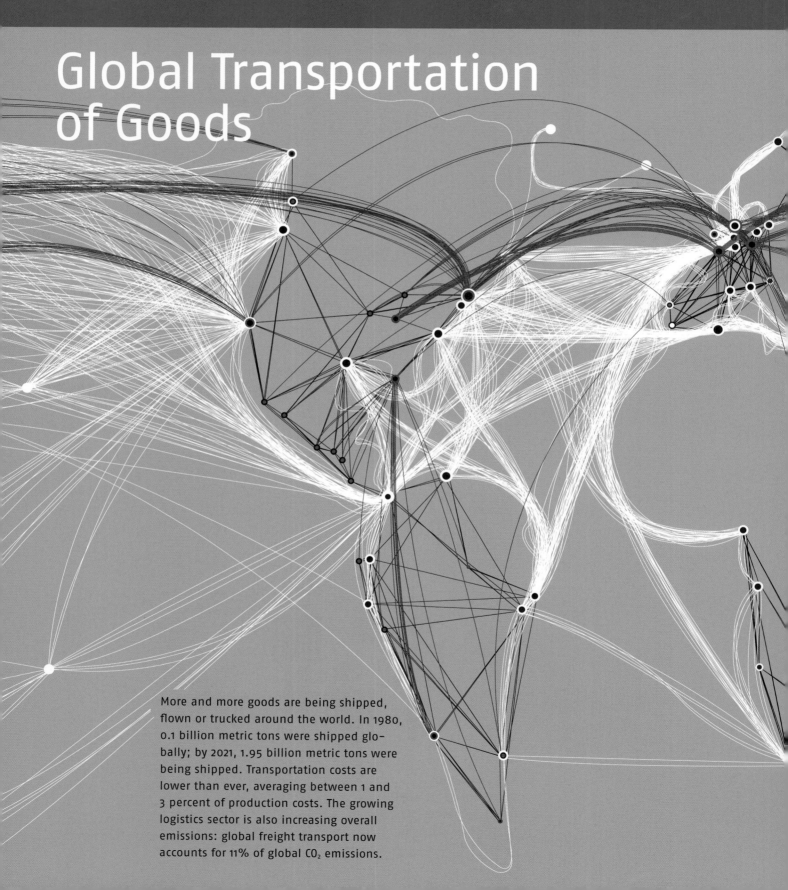

Global Transportation of Goods

More and more goods are being shipped, flown or trucked around the world. In 1980, 0.1 billion metric tons were shipped globally; by 2021, 1.95 billion metric tons were being shipped. Transportation costs are lower than ever, averaging between 1 and 3 percent of production costs. The growing logistics sector is also increasing overall emissions: global freight transport now accounts for 11% of global CO_2 emissions.

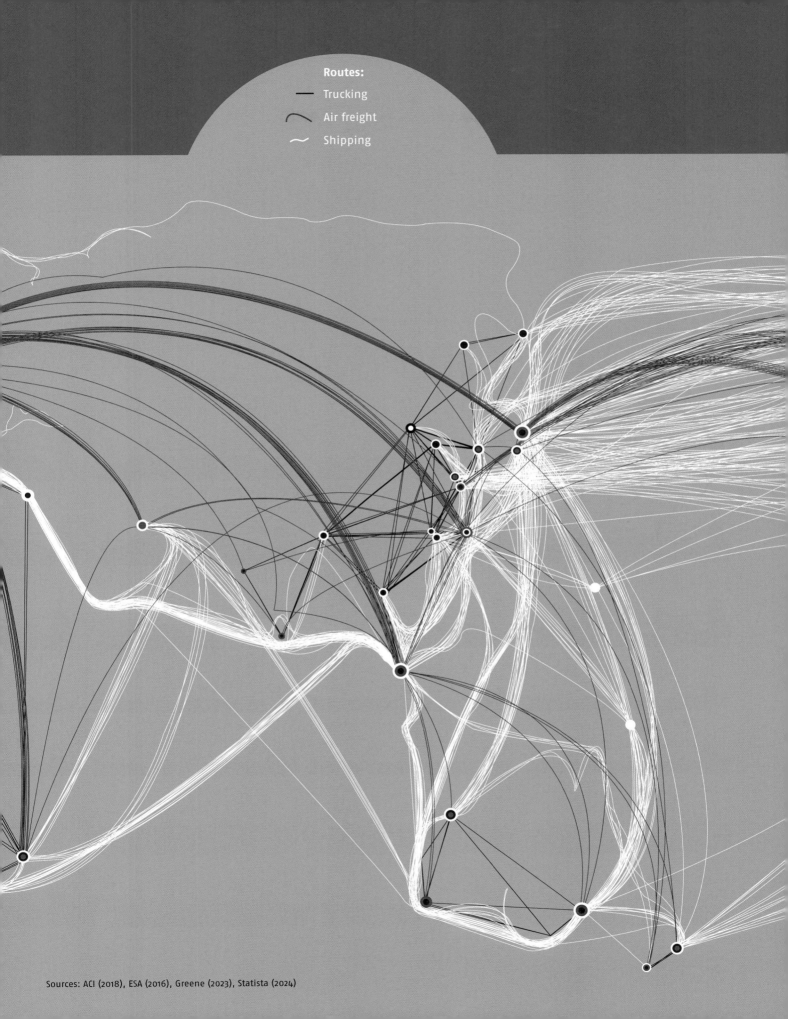

Transport by Sea

91 GT World merchant fleet 1955

2,062,000,000 dwt World merchant fleet 2020

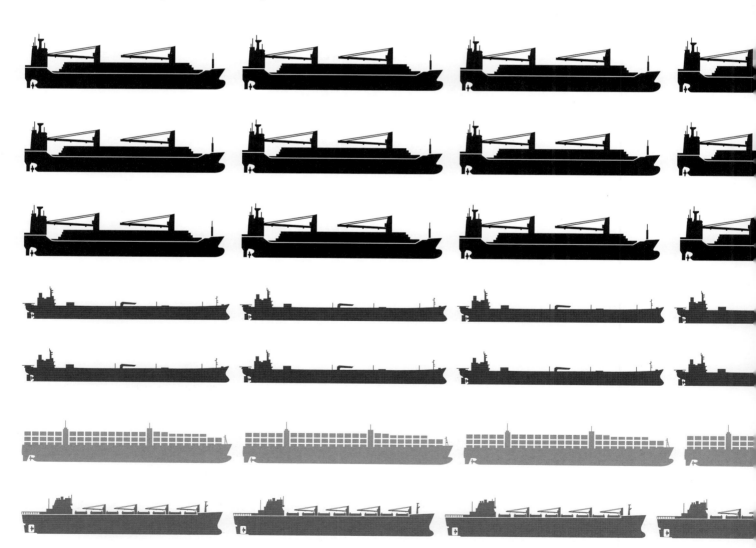

7,300,000 dwt Ferries and passenger ships

GT = Gross Tonnage
dwt = deadweight tonnage
(carrying capacity)

Around 80% of world trade is carried by some 62,100 cargo ships – they are the real engine of globalization. As the size of the ships increases, the shipping price per container goes down, but the total amount of emissions goes up.

Most cargo and cruise ships run on heavy fuel oil without soot filters. The resulting emissions of sulphur oxides, heavy metals, and particulate matter contribute significantly to health problems. This is particularly true in coastal areas, where the majority of emissions from shipping take place.

1955

2020

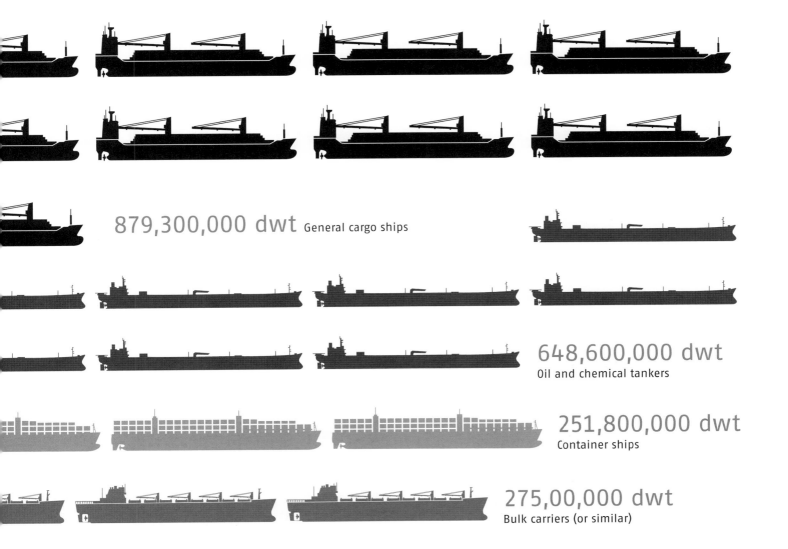

879,300,000 dwt General cargo ships

648,600,000 dwt
Oil and chemical tankers

251,800,000 dwt
Container ships

275,00,000 dwt
Bulk carriers (or similar)

Sources: KEG (1970), NABU (2014), UNCTAD (2020)

Transportation Routes on Land

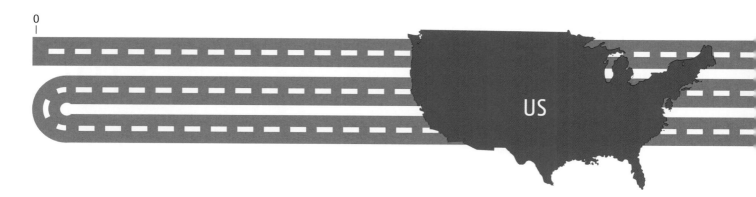

US

Number of registered motor vehicles
in the United States, 2021

GHG emissions of transportation
in the United States, 2021

58%
Cars

23%
Trucks

New road construction
vs. expansion of the transit network

in the United States, 2011–2019

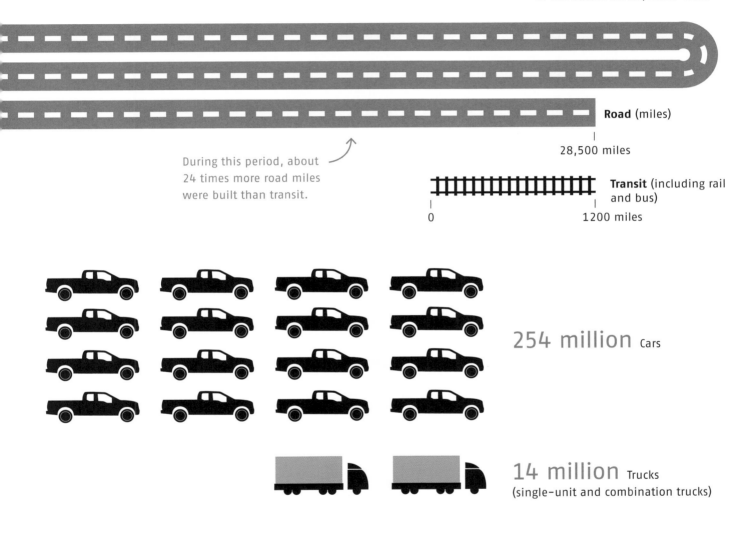

Road (miles)

28,500 miles

During this period, about 24 times more road miles were built than transit.

Transit (including rail and bus)

0   1200 miles

254 million Cars

14 million Trucks
(single-unit and combination trucks)

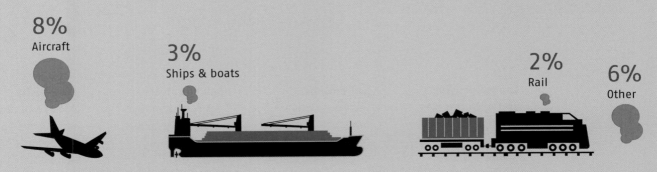

8%
Aircraft

3%
Ships & boats

2%
Rail

6%
Other

Sources: ATA (2022), DOT (2022), EPA (2023), Freemark (2020)

Infrastructure Drives Deforestation...

5%
of deforestation occurs more than 5 km away from any infrastructure.

95%
of deforestation occurs within a 5 km radius of roads or navigable rivers.

The first road is built through the forest.

Workers move into the region more and more.

Villages and small fields emerge.

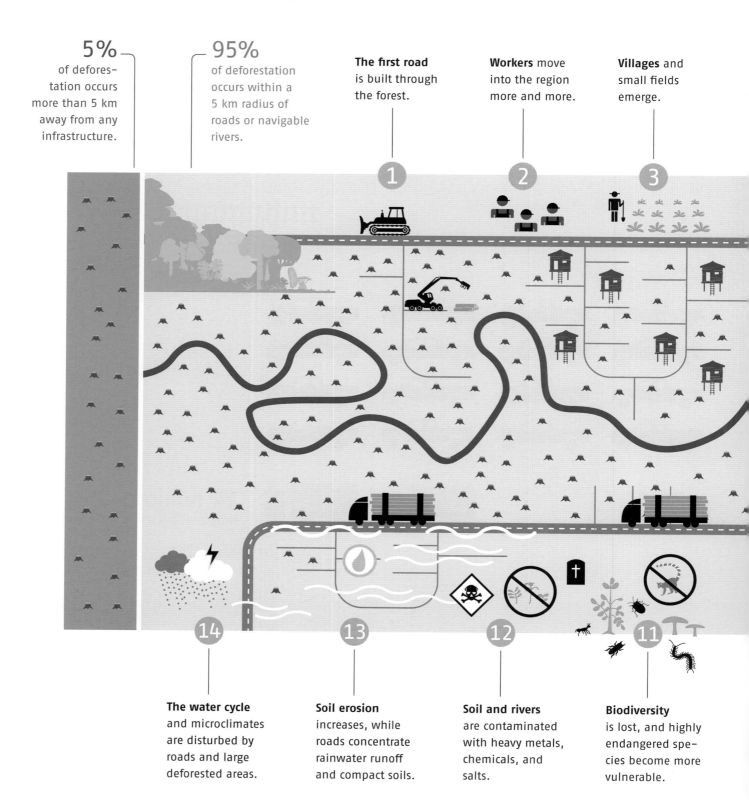

The water cycle and microclimates are disturbed by roads and large deforested areas.

Soil erosion increases, while roads concentrate rainwater runoff and compact soils.

Soil and rivers are contaminated with heavy metals, chemicals, and salts.

Biodiversity is lost, and highly endangered species become more vulnerable.

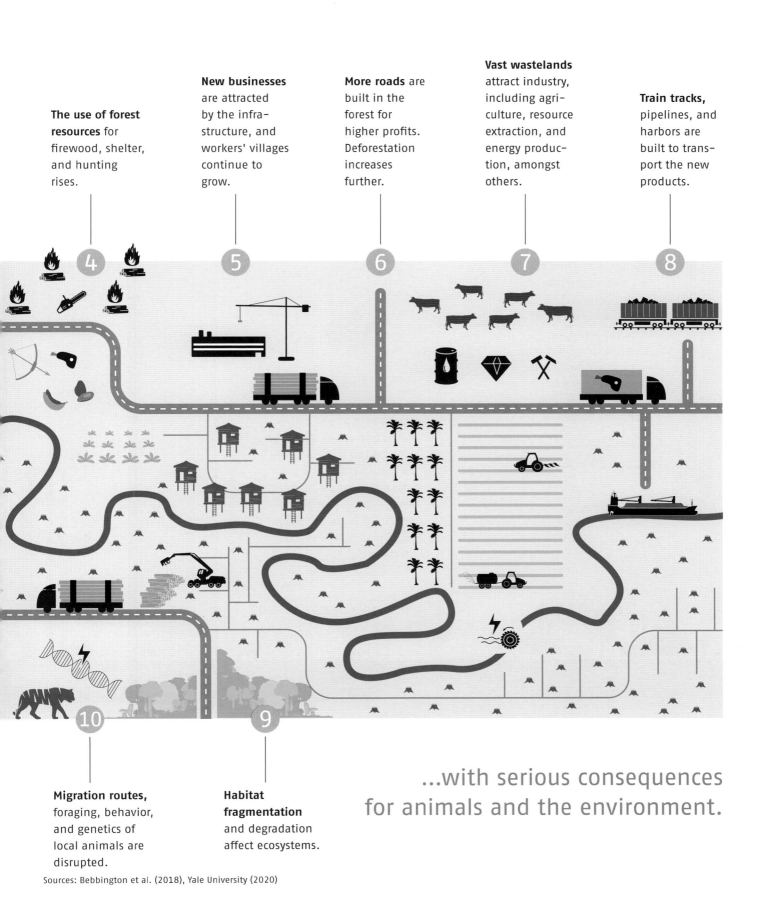

The use of forest resources for firewood, shelter, and hunting rises.

New businesses are attracted by the infrastructure, and workers' villages continue to grow.

More roads are built in the forest for higher profits. Deforestation increases further.

Vast wastelands attract industry, including agriculture, resource extraction, and energy production, amongst others.

Train tracks, pipelines, and harbors are built to transport the new products.

...with serious consequences for animals and the environment.

Migration routes, foraging, behavior, and genetics of local animals are disrupted.

Habitat fragmentation and degradation affect ecosystems.

Sources: Bebbington et al. (2018), Yale University (2020)

Green Cities & Urban Forests

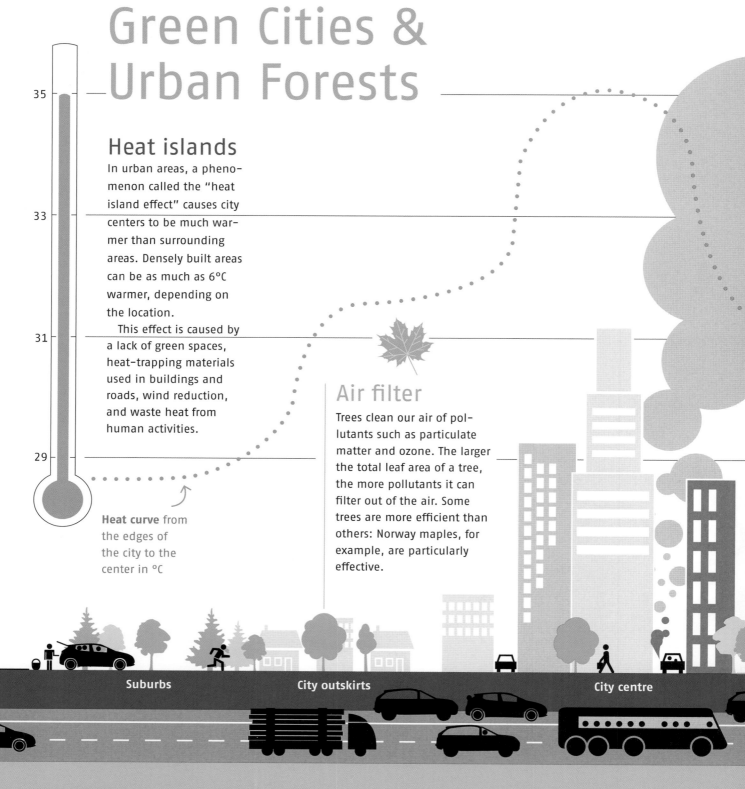

Heat islands

In urban areas, a pheno-menon called the "heat island effect" causes city centers to be much war-mer than surrounding areas. Densely built areas can be as much as 6°C warmer, depending on the location.

This effect is caused by a lack of green spaces, heat-trapping materials used in buildings and roads, wind reduction, and waste heat from human activities.

Heat curve from the edges of the city to the center in °C

Air filter

Trees clean our air of pol-lutants such as particulate matter and ozone. The larger the total leaf area of a tree, the more pollutants it can filter out of the air. Some trees are more efficient than others: Norway maples, for example, are particularly effective.

Suburbs   **City outskirts**   **City centre**

Heavy rain
Thanks to their roots, trees ensure better water drainage and seepage.

Air Temperature
The shade of a tree canopy can lower the surface and air temperature, as well as the indoor temperature of neighboring buildings.

Green cities with lots of trees and parks improve people's quality of life and health. In times of climate crisis, more heat waves are expected, and urban populations (which will account for about 60% of the world's population by 2030) will suffer more as cities warm at twice the rate of more rural areas.

Roads, sidewalks, and buildings absorb heat from the sun and warm their surroundings. Meanwhile, trees, plants, and lawns provide cooling through shade and evaporation. The temperature difference between shaded areas and sunny asphalt can be more than 20°C.

Trees cool the environment best when they have enough water: to keep the soil moist and the environment cool, urban trees are most effective when planted in the company of other plants and trees.

Natural A/C

On a hot summer's day, trees and green roofs can significantly cool their surroundings. By evaporating up to 400 liters of water, they convert heat into cool, fresh air.

Habitat

Squirrels, bats, bees, and songbirds: urban trees provide a habitat for many animals, increasing urban biodiversity and making cities more liveable.

Parks **Urban residential areas** **City outskirts** **Suburbs**

Noise
Trees and green facades reduce the noise intensity of road traffic.

Wind
Trees slow down gusts of wind and reduce the whirling up of dust during storms.

Sources: Livesley (2016), Marx (2017), Moser et al. (2017), Rahman et al. (2018), Roloff (2013), Winbourne et al. (2020)

Agriculture

How We Feed the World

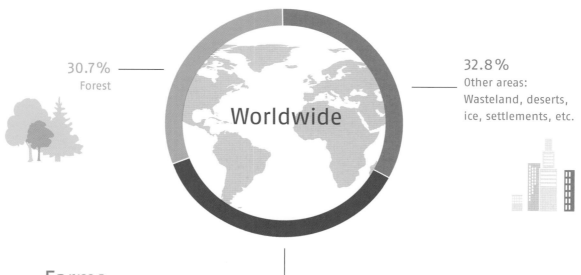

30.7%
Forest

Worldwide

32.8%
Other areas:
Wasteland, deserts,
ice, settlements, etc.

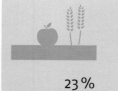

36.5%
Agricultural land,
broken down by use:

77%
Animal breeding, meat,
and dairy production

23%
Cultivation
of plants

Farms

More than

608 million

farms exist worldwide. 84% are
smaller than 2 hectares. More
than a third of the world's food
is grown on these small farms.

**Small farms suffer the most
from the climate crisis...**

Heatwaves

Droughts

Forest fires

Floods

Heavy rain and
severe weather

Diseases

Pollutants

2.66 Mt

of pesticides were sprayed on the world's fields in 2020. This could fill more than 1000 Olympic swimming pools.

x 1046

408,000 t

of pesticides were sprayed in the USA in 2020, and almost as much in Brazil (377,000 tonnes). China is in third place worldwide with 263,000 t.

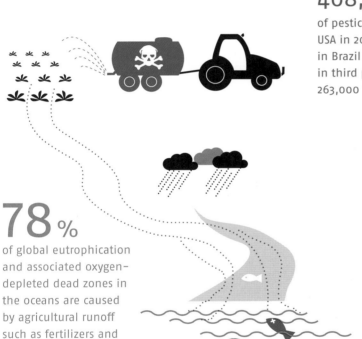

78%

of global eutrophication and associated oxygen-depleted dead zones in the oceans are caused by agricultural runoff such as fertilizers and pesticides.

Freshwater consumption

Worldwide

Households 11%

Industry 19%

At **70**%, agriculture is the largest consumer of water worldwide.

Productivity **+28**%

Global agriculture needs to become 28% more productive to meet the Paris Climate Agreement emissions targets and the United Nations goal of zero hunger by 2030.

Sources: FAO (2021), Fernandez (2022), Lowder (2021), OECD–FAO (2022), Sharma et al. (2019)

The Agricultural Transition

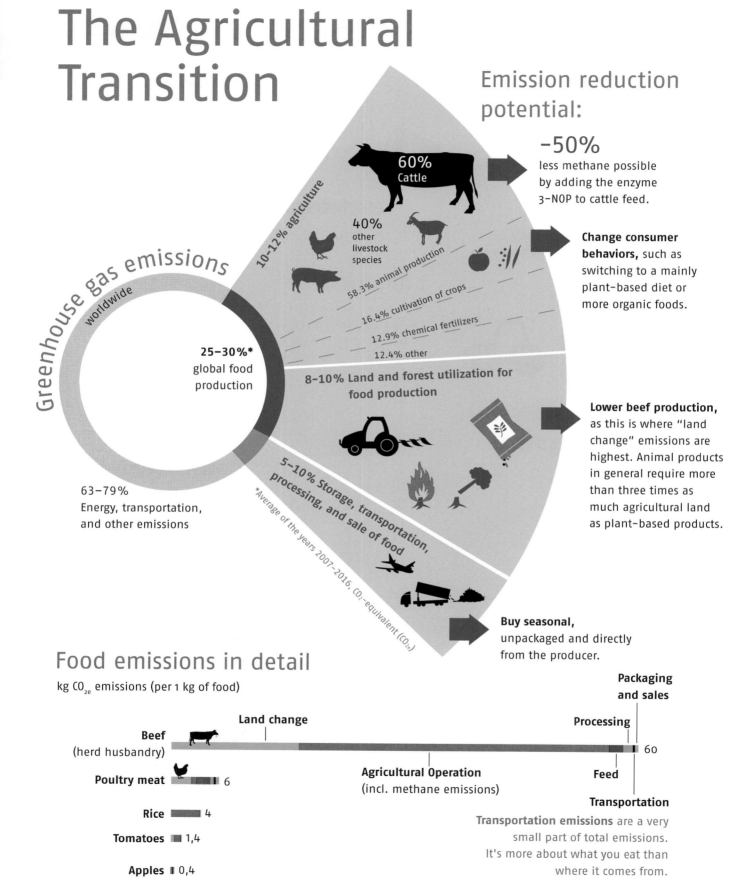

Emission reduction potential:

−50% less methane possible by adding the enzyme 3-NOP to cattle feed.

Change consumer behaviors, such as switching to a mainly plant-based diet or more organic foods.

Lower beef production, as this is where "land change" emissions are highest. Animal products in general require more than three times as much agricultural land as plant-based products.

Buy seasonal, unpackaged and directly from the producer.

Greenhouse gas emissions worldwide

25–30%* global food production

63–79% Energy, transportation, and other emissions

10–12% agriculture

60% Cattle

40% other livestock species

58.3% animal production

16.4% cultivation of crops

12.9% chemical fertilizers

12.4% other

8–10% Land and forest utilization for food production

5–10% Storage, transportation, processing, and sale of food

*Average of the years 2007–2016, CO₂-equivalent (CO₂e)

Food emissions in detail

kg CO₂e emissions (per 1 kg of food)

Beef (herd husbandry)

Poultry meat — 6

Rice — 4

Tomatoes — 1,4

Apples — 0,4

Land change

Packaging and sales

Processing

Agricultural Operation (incl. methane emissions)

Feed

Transportation

60

Transportation emissions are a very small part of total emissions. It's more about what you eat than where it comes from.

Sources: Gerber et al. (2013), Guégan & Léger (2015), IPCC (2019), Thornton et al. (2018), De Ramon N'Yeurt et al. (2012), Ritchie & Roser (2020)

Food security is jeopardized by the climate crisis

−20 %
lower crop yield by 2050 for varieties such as maize in Africa

Solution: Adapting farming techniques

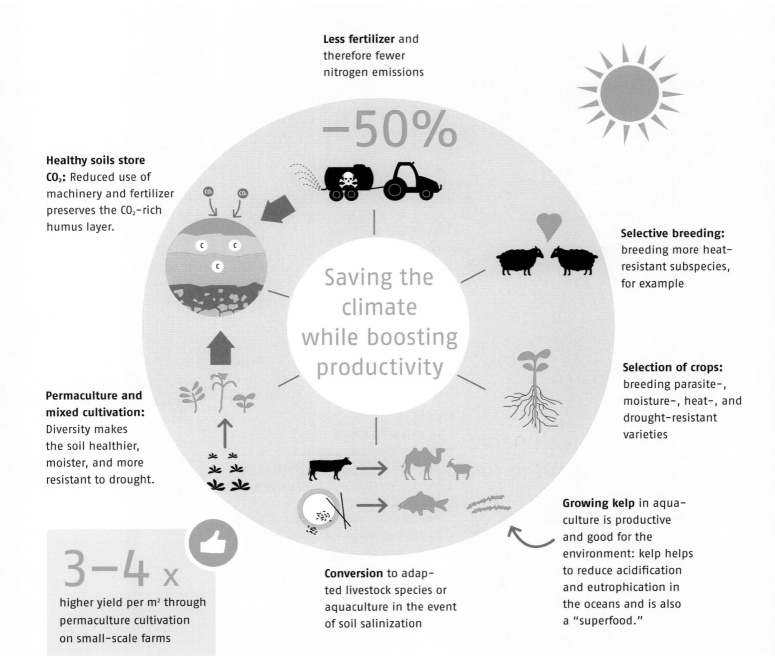

Less fertilizer and therefore fewer nitrogen emissions

−50 %

Healthy soils store CO₂: Reduced use of machinery and fertilizer preserves the CO_2-rich humus layer.

Selective breeding: breeding more heat-resistant subspecies, for example

Selection of crops: breeding parasite-, moisture-, heat-, and drought-resistant varieties

Permaculture and mixed cultivation: Diversity makes the soil healthier, moister, and more resistant to drought.

Saving the climate while boosting productivity

Growing kelp in aquaculture is productive and good for the environment: kelp helps to reduce acidification and eutrophication in the oceans and is also a "superfood."

Conversion to adapted livestock species or aquaculture in the event of soil salinization

3−4 x
higher yield per m² through permaculture cultivation on small-scale farms

Sustainable Farming – Agroforestry

1 Fields & trees
(Silvoarable systems)

Alley Cropping
Trees and crops are planted in parallel rows, sometimes so close together that the treetops can touch. Fruit trees or valuable timber are usually planted, such as walnut, cherry, oak, or chestnut. In between the rows, farmers grow maize, wheat, rye, potatoes, or beans.

Riparian Conservation Strips
Designed specifically around rivers and lakes, riparian strips may accommodate agroforestry products to stabilize the riparian zone, protect water quality, and increase biodiversity. In some countries and regions, they are even required by law on public and private land.

Windbreak Hedges
On the windward side of fields, usually in orchards, various shrubs are planted in U or L shapes in single or multiple rows. They not only create a calmer, cooler, and wetter microclimate in the field, but also provide a habitat for many species, increasing biodiversity.

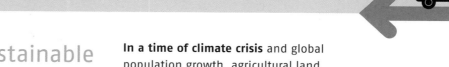

Sustainable & Adaptive

In a time of climate crisis and global population growth, agricultural land must not displace forests. Instead, we need to use the land we have in a smarter, more productive, and more versatile way. For example, by strategically combining agriculture with trees and forests. In the long term, not only farmers but society as a whole will benefit from the resulting environmental and economic improvements.

② Animals & trees
(Silvopastoral systems)

Semi-open pasture landscapes & orchards
Wooded pastures provide shade on hot days and increase soil moisture during droughts. Depending on the forest cover, firewood, mushrooms, berries, fodder, or bedding can also be harvested.

Forest Farming
Mainly practiced in the USA and Canada, forest farming involves growing valuable medicinal plants, herbs, mushrooms, or flowers in a section of forest, while simultaneously harvesting timber in a way that preserves the forest's climate.

③ Fields, animals & trees
(Agrosilvopastoral systems)

Dehesa & Forest Gardens
The "dehesa" system was introduced in Spain around 2500 BC. In this system, grazing land is also used to cultivate products like wood, forest gardens, fruit, vegetables, or honey.

Sources: Franklin et al. (2018), FAO (2020), DeFAF (2020)

Personal Change

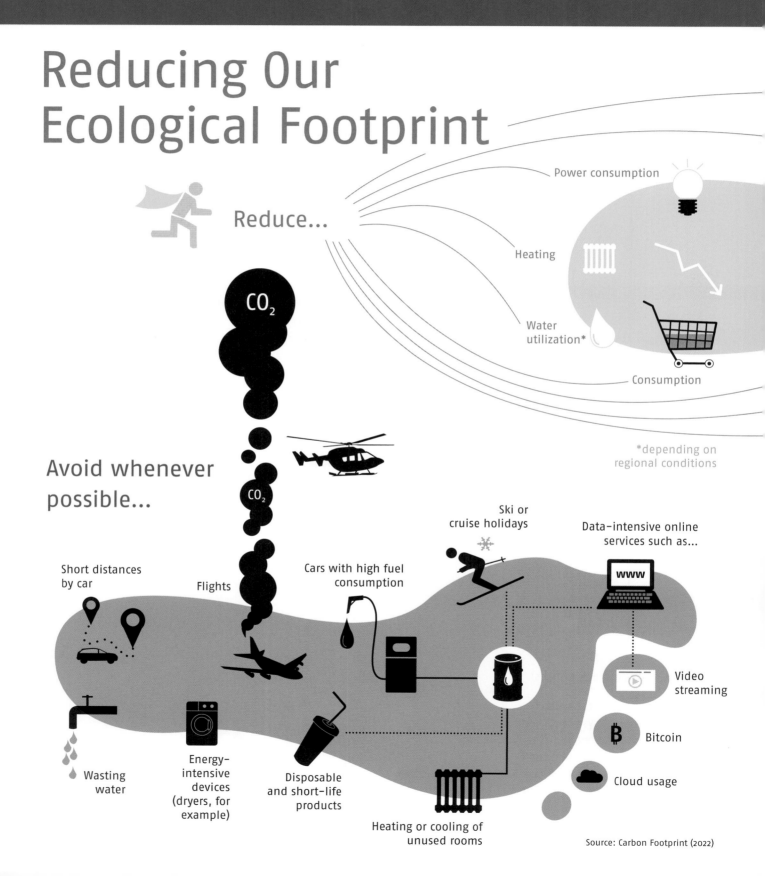

Reducing Our Ecological Footprint

Reduce...

Power consumption

Heating

Water utilization*

Consumption

*depending on regional conditions

CO₂

Avoid whenever possible...

CO₂

Short distances by car

Flights

Cars with high fuel consumption

Ski or cruise holidays

Data-intensive online services such as...

WWW

Video streaming

Bitcoin

Cloud usage

Wasting water

Energy-intensive devices (dryers, for example)

Disposable and short-life products

Heating or cooling of unused rooms

Source: Carbon Footprint (2022)

...and increase

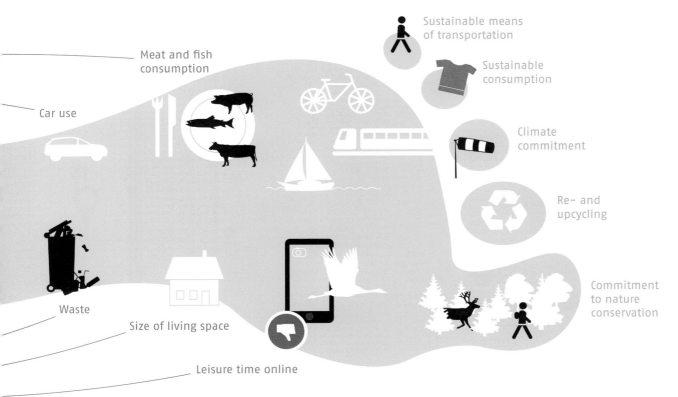

Meat and fish consumption

Car use

Sustainable means of transportation

Sustainable consumption

Climate commitment

Re- and upcycling

Commitment to nature conservation

Waste

Size of living space

Leisure time online

Have the courage to change...

Switch to an environmentally friendly bank and insurance company

Try cycling, hiking & canoeing trips

Use free climate training programs

Switch to green energy providers

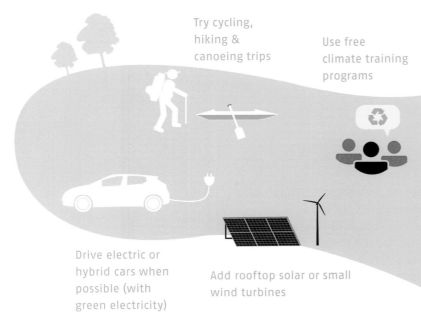

Drive electric or hybrid cars when possible (with green electricity)

Add rooftop solar or small wind turbines

Switch to an electric heat pump

Consuming Sustainably

Source local wood instead of tropical or even better: make use of recycled wood!

Reduce packaging waste

Aim for zero waste.

Buy seasonal food directly from local producers.

Avoid products containing palm oil where possible.

Favor plant-based and organic foods.

Reduce meat consumption to a minimum.

Improve your Climate Handprint!
A Climate Handprint is the opposite of a Carbon Footprint: it measures the positive impacts of our actions on the climate. This might be measured for an individual, company, or country. Where and how can our handprint grow?
You can find more inspiration here:
#ClimateHandprint

Source: KlimAktiv (2020)

Pay attention
to longevity
when buying
something new.

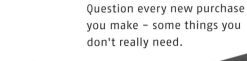
Don't

Question every new purchase
you make – some things you
don't really need.

BUY ME

Favor quality over
quantity, especially
when it comes
to clothing.

Buy used items
whenever possible –
people often sell items
in mint condition.

Buy fewer electrical
appliances and
look for manual
alternatives.

Repair instead of
buying new, perhaps
with the help of
"repair cafés."

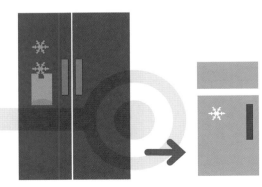

AAA+

Favor small and
energy-efficient
electrical appliances.

Protecting Nature

Anyone with ideas and drive can make a big difference.

On **209** weekends, lawyer Afroz Shah collected 7000 tonnes of garbage from Versova Beach in Mumbai. He was eventually joined by 220,000 volunteers! The UN then launched the global "Clean Seas" campaign.

13 mha of marine and coastal habitats have been placed under protection in Argentina. Pablo Garcia Borboroglu and his NGO "Global Penguin Society" have been working towards this goal for 25 years.

16-year-old John Abad from Peru, winner of the International Young Eco-Hero Award, mobilizes thousands of people to collect plastic. He also studies the pollution of the Chillón River, and presents his findings to the Minister of the Environment, among others.

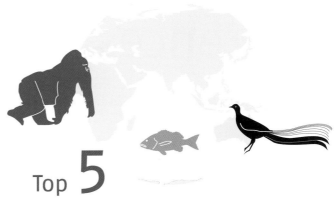

Top **5**

of the best-known conservationists: documentary filmmaker David Attenborough, gorilla expert Dr. Dian Fossey, marine biologist Dr. Sylvia Earle, chimpanzee researcher Dr. Jane Goodall, and climate activist Greta Thurnberg. These figures have brought global public attention to nature and climate protection.

108,000 USD

was collected by German nature photographer and marine biologist Robert Marc Lehmann during a YouTube livestream about shark protection. #haibockangriff

At **112** years old, environmentalist Saalumarada Thimmakka from India is still looking after the banyan forest she has planted over the past 70 years. Since 2022, she has held a cabinet rank for environmental protection.

Sources: Captain Planet Foundation (2022), Hindustan Times (2022), Lehmann (2022), Smithsonian (2022), UNEP (2016)

References All web links as of 3.15.2024

Used multiple times:

Intergovernmental Panel on Climate Change IPCC (2019): IPCC Special Report on the Ocean and Cryosphere in a Changing Climate. https://www.ipcc.ch/srocc

Intergovernmental Panel on Climate Change (IPCC) (2021): Climate Change 2021. The Physical Science Basis. Working Group I. https://www.ipcc.ch/report/ar6/wg1/

IPCC (WG-II) (2022): Climate Change 2022. Mitigation of Climate Change. Working Group II. https://www.ipcc.ch/report/sixth-assessment-report-working-group-ii/

IPCC (WG-III) (2022): Climate Change 2022: Impacts, Adaptation and Vulnerability. Working Group III. https://www.ipcc.ch/report/sixth-assessment-report-working-group-3/

World Ocean Review (WOR) (2019): World Ocean Review 6, The Arctic and Antarctic —Extreme, Climatically Crucial and In Crisis. Maribus Publishing, Hamburg. https://worldoceanreview.com/en/wor-6/

Pages	References
006 | 007	Aspen Global Change Institute (AGCI) (2022): Earth Systems. https://www.agci.org/earth-systems/geosphere
008	NOAA (2024): Layers of the Atmosphere. https://www.noaa.gov/jetstream/atmosphere/layers-of-atmosphere

010 | 011   Markart, G.; Kohl, B. (2009): Wie viel Wasser speichert der Waldboden? Abflussverhalten und Erosion. BFW-Praxisinformation 19, 25-26. https://www.waldwissen.net/wald/schutzfunktion/wasser/bfw_wasserspeicher_boden/index_DE

Perlman, H. et al. (2019): The World's Water. https://www.usgs.gov/media/images/all-earths-water-a-single-sphere

United States Geological Survey (USGS) (2019): How Much Water is There on Earth? https://www.usgs.gov/special-topics/water-science-school/science/how-much-water-there-earth

Zimmermann, L. et al. (2008): Wasserverbrauch von Wäldern. https://www.lwf.bayern.de/mam/cms04/boden-klima/dateien/a66-wasserverbrauch-von-waeldern.pdf

012   Deutscher Wetter Dienst (DWD) (2022): Wetter- und Klimalexikon – Hochdruckgebiet. https://www.dwd.de/DE/service/lexikon/Functions/glossar.html?lv3=101176&lv2=101094; Wetter- und Klimalexikon – Tiefdruckgebiet. https://www.dwd.de/DE/service/lexikon/Functions/glossar.html?lv3=102762&lv2=102672

013   Coumou, D. et al. (2018): The influence of Arctic amplification on mid-latitude summer circulation. Nature Communications. https://www.nature.com/articles/s41467-018-05256-8

Kornhuber, K. et al. (2019): Amplified Rossby waves enhance risk of concurrent heatwaves in major breadbasket regions. Nature Climate Change. https://www.nature.com/articles/s41558-019-0637-z (12.08.2021)

National Oceanic and Atmospheric Administration (NOAA) (2019): The science behind the polar vortex. You might want to put on a sweater. https://www.noaa.gov/multimedia/infographic/science-behind-polar-vortex-you-might-want-to-put-on-sweater

NOAA (2021): How does sea ice affect global climate? Sea ice affects both global ocean temperatures and the global movement of ocean waters. https://oceanservice.noaa.gov/facts/sea-ice-climate.html

Voosen, P. (2020): New feedbacks speed up the demise of Arctic sea ice. Science 28.08.2020: Vol. 369, Ausgabe 6507. https://science.sciencemag.org/content/369/6507/1043

014 | 015   NOAA (2021): Saffir-Simpson Hurricane Wind Scale. National Hurricane Center. https://www.nhc.noaa.gov/aboutsshws.php; How do hurricanes form? https://oceanservice.noaa.gov/facts/how-hurricanes-form.html

Rahmstorf et al. (2018): Does global warming make tropical cyclones stronger?

https://www.realclimate.org/index.php/archives/2018/05/does-global-warming-make-tropical-cyclones-stronger/

Velden C. et al. (2017): Reprocessing the Most Intense Historical Tropical Cyclones in the Satellite Era Using the Advanced Dvorak Technique. Mon. Weather Rev. 145(3):971–983.

016–019 DWD (2018): Klimawandel – ein Überblick.
https://www.dwd.de/DE/klimaumwelt/klimawandel/ueberblick/ueberblick_node.html

Rahmstorf, S. (2013): Wie funktioniert eigentlich der Treibhauseffekt? http://www.pik-potsdam.de/~stefan/leser_antworten.html

Riedel, E., Janiak, C. (2015): Anorganische Chemie. De Gruyter Studium.

United Nations Climate Change (UNFCCC) (2022): What is the Kyoto Protocol? https://unfccc.int/kyoto_protocol

024 | 025 NOAA (2022): Greenhouse gas emissions are on the rise. https://research.noaa.gov/article/ArtMID/587/ArticleID/2816

UN (2013): World Population Prospects: The 2012 Revision. New York.
https://population.un.org/wpp/Publications/Files/WPP2012_highlights.pdf

UN (2018): World Population Prospects 2017. https://population.un.org/wpp/Graphs/

026 | 027 Global Carbon Atlas (GCA) (2022): Fossil Fuel Emissions. Interaktive Karte. http://www.globalcarbonatlas.org/en/CO2-emissions

028 | 029 Loeb, N. G. et al. (2021): Satellite and Ocean Data Reveal Marked Increase in Earth's Heating Rate. Geophysical Research Letters.
https://agupubs.onlinelibrary.wiley.com/doi/10.1029/2021GL093047

National Aeronautics and Space Administration (NASA) (2022): GISS Surface Temperature Analysis (v4). Global Maps.
https://data.giss.nasa.gov/gistemp/maps/

NOAA (2022): Climate at a Glance. Global Time Series.
https://www.ncei.noaa.gov/access/monitoring/climate-at-a-glance/global/time-series

030 | 031 Cheng, L. et al. (2019): How fast are the oceans warming? Science, 11.01.2019: Vol. 363,
Issue 6423, DOI: 10.1126/science.aav7619. https://science.sciencemag.org/content/363/6423/128

EPA (2016): Climate Change Indicators in the United States: Ocean Heat.
https://www.epa.gov/climate-indicators/downloads-indicators-report

Gleckler, P. J. et al. (2016): Industrial-era global ocean heat uptake doubles in recent decades. Nature Climate Change volume 6,
p. 394–398. https://www.nature.com/articles/nclimate2915

Resplandy, L. et al. (2018): Quantification of ocean heat uptake from changes in atmospheric O_2 and CO_2 composition.
Nature 563, 105–108. https://www.nature.com/articles/s41586-018-0651-8

032 | 033 Armstrong McKay, D. I. et al. (2022): Exceeding 1.5°C global warming could triggerbmultiple climate tipping points.
Science Volume 377, Issue 6611, Sep 2022. https://www.science.org/doi/epdf/10.1126/science.abn7950

034 | 035 Cho, R. (2021): Climate Migration: An Impending Global Challenge.
https://news.climate.columbia.edu/2021/05/13/climate-migration-an-impending-global-challenge/

036 | 037 Carbon Brief (CB) (2017): Mapped: How climate change affects extreme weather around the world.
https://www.carbonbrief.org/mapped-how-climate-change-affects-extreme-weather-around-the-world

Funk, C. et al. (2015): Bull. Amer. Meteor. Soc., 96(12), p. 77–82.

Kam, J. et al. (2015): Bull. Amer. Meteor. Soc., 96(12), p. 61–65.

King, A. D. et al. (2015): Environmental Research Letters, 10(5), 54002.

Junghänel, T. et al. (2021): Hydro-klimatologische Einordnung der Stark- und Dauerniederschläge in Teilen Deutschlands im Zusammenhang mit dem Tiefdruckgebiet „Bernd" vom 12. bis 19. Juli 2021. Deutscher Wetterdienst.

References

https://rcc.dwd.de/DE/leistungen/besondereereignisse/niederschlag/20210721_bericht_starkniederschlaege_tief_bernd.html

Murakami, H. et al. (2015): Bull. Amer. Meteor. Soc., 96(12), p. 115–119.

Shiogama, H. et al. (2013): Atmospheric Science Letters, 14(3), p. 170–175.

Sweet, W. V. et al. (2013): Bull. Amer. Meteor. Soc., 97(12), p. 25–30.

Sweet, W. V. et al. (2013): Bull. Amer. Meteor. Soc., 94(9), p. 17–20.

Szeto, K. et al. (2016): Bull. Amer. Meteor. Soc., 97(12), p. 42–46.

Zhang, W. et al. (2016): Bull. Amer. Meteor. Soc., 97(12), p. 131–135.

World Weather Attribution (WWA) (2020): Attribution of the Australian bushfire risk to anthropogenic climate change. https://www.worldweatherattribution.org/bushfires-in-australia-2019-2020/

WWA (2021): Siberian heatwave of 2020 almost impossible without climate change. https://www.worldweatherattribution.org/siberian-heatwave-of-2020-almost-impossible-without-climate-change/

WWA (2021): Climate Change made devastating early heat in India and Pakistan 30 times more likely. https://www.world weatherattribution.org/climate-change-made-devastating-early-heat-in-india-and-pakistan-30-times-more-likely/

038 | 039   Eakin, C. M. et al. (2018): Unprecedented three years of global coral bleaching 2014–17.
Bulletin of the American Meteorological Society, 99(8), p. 74–S75.

Friedlingstein, P. et al. (2022): https://www.globalcarbonproject.org/carbonbudget/22/files/GCP_CarbonBudget_2022.pdf
MCC (2022): https://www.mcc-berlin.net/en/research/co2-budget.html

Mercator Research Institute on Global Commons and Climate Change (MCC) (2022): That's how fast the Carbon Clock is ticking. https://www.mcc-berlin.net/en/research/co2-budget.html

Mora, C. et al. (2017): Global risk of deadly heat. Nature Climate Change volume 7, pages 501–506 (2017)
https://www.nature.com/articles/nclimate3322 und https://maps.esri.com/globalriskofdeadlyheat

NASA (2022): Sea Level Rise Viewer. Interaktive Karte. https://sealevel.nasa.gov/data_tools/17

World Ressource Institute (WRI) (2022): Reefs at risk revisited. Interaktive Karte.
https://www.wri.org/our-work/project/reefs-risk/interactive-map#project-tabs

040 | 041   Bai, X. et al. (2018): Six research priorities for cities and climate change. Nature 555, 23-25 (2018).
DOI: 10.1038/d41586-018-02409-z. https://www.nature.com/articles/d41586-018-02409-z

Gitz, V. et al. (2016): Climate change and food security: risks and responses.
Food and Agriculture Organization of the United Nations. http://www.fao.org/3/a-i5188e.pdf

Scherer, M., Tänzler, D. (2018): The Vulnerable Twenty – From Climate Risks to Adaptation. A compendium of climate fragility risks and adaptation finance needs of the V20 countries. adelphi research.
https://www.adelphi.de/en/publication/vulnerable-twenty

042 | 043   Irvine, P., Keith, D. (2021): The US Can't Go It Alone on Solar Geoengineering. Environmental Affaires.
https://policyexchange.org.uk/the-us-cant-go-it-alone-on-solar-geoengineering/

044 | 045   BP (2022): Statistical Review of World Energy.
https://www.bp.com/en/global/corporate/energy-economics/statistical-review-of-world-energy.html

International Renewable Energy Agency (Irena) (2021): IRENA (2021). Renewable Energy and Jobs – Annual Review 2021.
https://irena.org/publications/2021/Oct/Renewable-Energy-and-Jobs-Annual-Review-2021
International Solar Alliance (ISA) (2022): World Solar Investment Report.
https://isolaralliance.org/uploads/docs/a15941394105403ec59b9b1e569c7d.pdf

The World Bank (WBG) (2017): Annual Report 2017.
http://pubdocs.worldbank.org/en/908481507403754670/Annual-Report-2017-WBG.pdf

World Health Organization (WHO) (2018): Interactive global ambient air pollution map. https://www.who.int/airpollution/en/

046   Quaschning, V. (2021): Erneuerbare Energien und Klimaschutz. Carl Hanser Verlag München.

Rollet, C. (2019): Are rare earths used in solar panels?
https://www.pv-magazine.com/2019/11/28/are-rare-earths-used-in-solar-panels/

047   Hornung, C. (2020): Windenergien. Wie funktioniert eine Windkraftanlage? enercity Magazin.
https://www.enercity.de/magazin/unsere-welt/so-funktioniert-eine-windkraftanlage

Fechner, H., Zwiauer, K. (2021): Grundlagen Windenergie. Geschichte der Windenergie.
https://www.e-genius.at/lernfelder/erneuerbare-energien/grundlagen-windenergie

048 | 049   Bundesverband Windenergie (BWE) (2021): Windenergie im Forst.
https://www.wind-energie.de/themen/mensch-und-umwelt/wind-im-forst/

Ember (2022): Global Electricity Review 2022. https://ember-climate.org/insights/research/global-electricity-review-2022

International Energy Agency (IEA) (2012): Net Zero by 2050. Report. https://www.iea.org/reports/net-zero-by-2050

Kühl, A. (2019): 24 Fakten zur Photovoltaik. Solarstrom Magazin Solarimo.
https://solarimo.de/solarstrom-magazin/24-fakten-photovoltaik/

050 | 051   IEA (2022): Transport sector CO_2 emissions by mode in the Sustainable Development Scenario, 2000-2030. https://www.iea.org/data-and-statistics/charts/transport-sector-co2-emissions-by-mode-in-the-sustainable-development-scenario-2000-2030

Paoli, L. et al. (2021): Electric Vehicles. IEA Report. https://www.iea.org/reports/electric-vehicles

Tiseo, I. (2021): Greenhouse gas emissions from the transportation sector worldwide from 1990 to 2018. Statista.
https://www.statista.com/statistics/1084096/ghg-emissions-transportation-sector-globally/#:~:text=Global%20greenhouse%20emissions%20from%20the,during%20the%202009%20global%20recession

Plötz, P., Link, S. (2021): Projekt Zero-Emission Delivery – Lieferverkehr mit Batterie-Lkw: Machbarkeit 2021. Fraunhofer-Institut für System- und Innovationsforschung ISI. https://www.transportenvironment.org wp-content/uploads/2021/12/2021_12_13_Fraunhofer-ISI_Webinar_Kann-der-Lieferverkehr-elektrisch.pdf

052 | 053   Carbfix (2022): We turn CO_2 into Stone. https://www.carbfix.com/

EnergyNow Media (2021): Introduction to Carbon Capture and Storage (CCS) Technology.
https://energynow.ca/2022/02/introduction-to-carbon-capture-and-storage-ccs-technology

Geoengineering Monitor (2021): Bioenergie mit CO_2-Abscheidung und –speicherung (BECCS).
https://www.boell.de/sites/default/files/2021-01/GM_BECCS_de.pdf

GEOMAR (2021): Den Ozean zum Verbündeten beim Klimaschutz machen.
https://www.geomar.de/news/article/den-ozean-zum-verbuendeten-beim-klimaschutz-machen

Mengis, N., Kalhori, A. (2021): Moore als natürliche CO_2-Speicher.
https://www.wissenschaftsjahr.de/2020-21/koepfe-des-wandels/moore-als-natuerliche-co2-speicher

UC Davis (2019): What is Carbon Sequestration and How Does it Work?
https://clear.ucdavis.edu/explainers/what-carbon-sequestration

054 | 055   IPCC (2014): Synthesis Report, Fifth Assessment Report. https://www.ipcc.ch/report/ar5/syr/

UNFCCC (2017): 20 Years of Effort and Achievement: http://unfccc.int/timeline/ (20.05.2019)
Adoption of the Paris Agreement: http://unfccc.int/resource/docs/2015/cop21/eng/l09r01.pdf

References

056 | 057   Burns, L. et al. (2019): Solar Geoengineering. Technology Factsheet Series. Harvard University.
https://geoengineering.environment.harvard.edu/geoengineering

Hüttmann, M. (2020): Geoengineering-Technologien: 9. Marine Cloud Brightening. Deutsche Gesellschaft für Sonnenenergie e. V.
https://www.dgs.de/news/en-detail/100120-geoengineering-technologien-9-marine-cloud-brightening/

Hüttmann, M. (2019): Geoengineering-Technologien: 2. Stratospheric Aerosol Injection.
https://www.dgs.de/news/en-detail/050719-geoengineering-technologien-2-stratospheric-aerosol-injection/

Irvine, P., Keith, D. (2020): Halving warming with stratospheric aerosol geoengineering moderates policy-relevant climate hazards.
Environ. Res. Lett. 15 044011. https://iopscience.iop.org/article/10.1088/1748-9326/ab76de#:~:text=Stratospheric%20aerosol%20
geoengineering%20is%20a,of%20reducing%20average%20climate%20changes.w

Irvine, P. (2022): Could solar geoengineering have a role in future climate policy? CfCA and Cross Government Climate Hub.
https://www.youtube.com/watch?v=cgaB5VS-oOw

060 | 061   Gartner, J., Armstrong, A. (2012): Mariana Trench and West Mariana Ridge, Pacific Ocean. The University of New Hampshire und
NOAA, Center for Coastal and Ocean Mapping Joint hydrographic Center.
http://ccom.unh.edu/jim-gardner-and-andy-armstrong-survey-mariana-trench

GEOMAR (2021): 10 Fakten über die Meere. https://www.geomar.de/en/news/article/10-fakten-ueber-die-meere

Mora, C. et al. (2011): How Many Species Are There on Earth and in the Ocean?
https://www.ncbi.nlm.nih.gov/pmc/articles/PMC3160336/

NOAA (2021): What are the oldest living animals in the world? https://oceanservice.noaa.gov/facts/oldest-living-animal.html

062 | 063   Heinrich-Böll-Stiftung (HBS) (2017): Meeresatlas 2017. Daten und Fakten über unseren Umgang mit dem Ozean.
https://www.boell.de/de/2017/04/25/meeresatlas-daten-und-fakten-ueber-unseren-umgang-mit-dem-ozean

064   ARC Centre of Excellence, Coral Reef Studies (ARC) (2022): Coral bleaching and the Great Barrier Reef
https://www.coralcoe.org.au/for-managers/coral-bleaching-and-the-great-barrier-reef

NOAA (2019): Coral reef ecosystems. https://www.noaa.gov/education/resource-collections/marine-life/
coral-reef-ecosystems#:~:text=Because%20of%20the%20diversity%20of,and%20crannies%20formed%20by%20corals.

065   IGBP, IOC, SCOR (2013): Ozeanversauerung. Zusammenfassung für Entscheidungsträger Third Symposium on the Ocean in a High-CO$_2$
World. http://www.igbp.net/download/18.2fc4e526146d4c130b72cf/1411549163212/OzeanversauerungZfE.pdf

Maribus (2010): World Ocean Review – Mit den Meeren leben. http://worldoceanreview.com

NOAA (2020): Ocean acidification. https://www.noaa.gov/education/resource-collections/ocean-coasts/ocean-acidification

066 | 067   Eriksen, M. et al. (2014): Plastic Pollution in the World's Oceans: More than 5 Trillion Plastic Pieces Weighing over 250,000 Tons
Afloat at Sea. PLoS ONE 9(12): e111913. doi:10.1371/journal.pone.0111913

Greenpeace (GP) (2007): Plastic Debris in the World's Oceans. http://www.greenpeace.org/international/Global/international
/planet-2/report/2007/8/plastic_ocean_report.pdf

Moore, C. J. et al. (2001): A Comparison of Plastic and Plakton in the North Pacific Central Gyre. Marine Bulletin 42 (12) 1297-1300.
http://www.sciencedirect.com/science/article/pii/S0025326X0100114X

Ocean Conservancy, International Coastel Cleanup (ICC) (2010): Trash Travels.
http://act.oceanconservancy.org/images/2010ICCReportRelease_pressPhotos/2010_ICC_Report.pdf

Maribus (2010): World Ocean Review – Mit den Meeren leben. http://worldoceanreview.com

United Nations Environment Programme (UNEP) (2005): Marine Litter. An Analytical Overview.
http://www.cep.unep.org/content/about-cep/amep/marine-litter-an-analytical-overview/view

068 | 069   Global Water System Project (GWSP) (2015): GRanD Database. http://sedac.ciesin.columbia.edu/data/collection/grand-v1

Lehner, B. et al. (2011): Global Reservoir and Dam Database, Version 1 (GRanDv1): Reservoirs, Revision 01. Palisades, NY: NASA Socioeconomic Data and Applications Center (SEDAC). http://dx.doi.org/10.7927/H4HH6H08.

070 | 071   Gustafson, C. et al. (2019): Aquifer systems extending far offshore on the U.S. Atlantic margin.
https://www.nature.com/articles/s41598-019-44611-7

Li, P. et al. (2021): Sources and Consequences of Groundwater Contamination.
https://link.springer.com/article/10.1007/s00244-020-00805-z

United States Geological Survey (USGS) (2018): Total Water Use in the United States.
https://www.usgs.gov/special-topics/water-science-school/science/total-water-use-united-states

072 | 073   Dahl, T.E. (2006): Status and Trends of Wetlands in the Conterminous United States 1998 to 2004. Washington, D.C.: U.S. Department of the Interior, Fish and Wildlife Service, 2006. https://www.fws.gov/sites/default/files/documents/Status-and-Trends-of-Wetlands-in-the-Conterminous-United-States-1998-to-2004.pdf.

Hooijer, A. et al. (2010): Current and future CO_2 emissions from drained peatlands in Southeast Asia.
https://helda.helsinki.fi/bitstream/handle/10138/29039/bg_7_1505_2010.pdf?sequence=2

Joosten, H. (2022): Moor muss nass. Wiedervernässung vorantreiben, Torfabbau verhindern, in: Wiegandt, K. (Hrsg.): 3 Grad mehr, S. 209–232, oekom verlag.

Page, S. E., Rieley, J. O., Banks, C. J. (2011). Global and regional importance of the tropical peatland carbon pool. Global Change Biology, 17(2), 798-818. https://doi.org/10.1111/j.1365-2486.2010.02279.x

074 | 075   Armstrong, R. L. et al. (2019): State of the Cryosphere. NSIDC. https://nsidc.org/cryosphere/sotc

National Oceanic and Atmospheric Administration (NOAA) (2021): Global Multisensor Snow/Ice Cover Map.
https://www.star.nesdis.noaa.gov/smcd/emb/snow/HTML/multisensor_global_snow_ice.html

SSEC (2017): The Cryosphere in RealEarth. SSEC/CIMSS, University of Wisconsin-Madison. https://re.ssec.wisc.edu/s/l1I6Y9

Wadhams, P. (2017): A Farewell to Ice. Penguin Books.

World Meteorological Organization (WMO) (2021): Global Cryosphere Watch. About the Cryosphere.
https://globalcryospherewatch.org/about/cryosphere.html

076 | 077   Garthwaite, J. (2019): Polar vortex. The science behind the cold. Stanford University.
https://earth.stanford.edu/news/polar-vortex-science-behind-cold#gs.ro4cog

Kornhuber, K. et al. (2019): Amplified Rossby waves enhance risk of concurrent heatwaves inmajor breadbasket regions.
Nature Climate Change. https://www.nature.com/articles/s41558-019-0637-z

National Oceanic and Atmospheric Administration (NOAA) (2021): The science behind the polar vortex.
https://www.noaa.gov/multimedia/infographic/science-behind-polar-vortex-you-might-want-to-put-on-sweater

Pruppacher, H. R., Klett, J. D. (1978): Microphysics of Clouds and Precipitation. D. Reidel Publishing Company, Boston, S. 714.

078 | 079   Alfred-Wegener-Institut (AWI) (2015): Das aktuelle Wissen zum Thema. Meereis. Fact Sheet. https://www.meereisportal.de/file admin/user_upload/www.meereisportal.de/MeereisPortal/Archiv_Kurzmeldungen/AWI_Factsheet_Meereis.pdf

080 | 081   Alfred-Wegener-Institut (AWI) (2020): Arktis und Antarktis – mehr Unterschiede als Gemeinsamkeiten? Fact Sheet.

082 | 083   Wang, X. et al. (2019): A new look at roles of the cryosphere in sustainable development. Advances in Climate Change Research 10 (2019) 124e131. https://www.sciencedirect.com/science/article/pii/S1674927818301084

084 | 085   Global Cryosphere Watch (GCW) (2021): About Glaciers and Ice Caps & About Freshwater Ice. World Meteorological Organization.
https://globalcryospherewatch.org/about.html
Intergovernmental Panel on Climate Change (IPCC) (2013): IPCC AR5 Working Group I. Polar Regions Polar Amplification, Permafrost, Sea ice changes. https://www.ipcc.ch/site/assets/uploads/2018/02/WG1AR5_SPM_FINAL.pdf

References

Slater, T. et al. (2021): Earth's ice imbalance. The Cryosphere, 15, S. 233–246. https://doi.org/10.5194/tc-15-233-2021

National Snow and Ice Data Center (NSIDC) (2021): State of the Cryosphere. Ice Sheets. https://nsidc.org/cryosphere/sotc/ice_sheets.html

086   Schuckmann, K. von et al. (2020): Heat stored in the Earth system. Where does the energy go?, Earth Syst. Sci. Data, Volume 12, Issue 3, 2020 https://doi.org/10.5194/essd-12-2013-2020

087   Siegert, M. et al. (2020): Twenty-first century sea-level rise could exceed IPCC projections for strong-warming futures. https://doi.org/10.1016/j.oneear.2020.11.002

Slater, T. et al. (2021): Earth's ice imbalance. The Cryosphere, 15, p. 233–246. https://doi.org/10.5194/tc-15-233-2021

088 | 089   Damania, R. et al. (2017): Uncharted Waters. The New Economics of Water Scarcity and Variability. World Bank. https://openknowledge.worldbank.org/handle/10986/28096

Luo, T. et al. (2015): Aqueduct Projected Water Stress Country Rankings. Technical Note. World Resources Institute. https://www.wri.org/publication/aqueduct-projected-water-stress-country-rankings

Hoekstra, A. (2008): Water for Food. The water footprint of food, S. 54. https://waterfootprint.org/media/downloads/Hoekstra-2008-WaterfootprintFood_1.pdf

Mekonnen, M., Hoekstra, A. (2016): Four Billion People Facing Severe Water Scarcity. Science Advances 2 (2), e1500323. http://advances.sciencemag.org/content/2/2/e1500323

Union of Concerned Scientists (UCS) (2011): Nuclear Power and Water. Quick facts on nuclear power generation and water use. https://www.ucsusa.org/sites/default/files/legacy/assets/documents/nuclear_power/fact-sheet-water-use.pdf

United Nations World Water Assessment Programme/UN-Water (WWAP) (2018): The United Nations World Water Development Report 2018. Nature-Based Solutions for Water. Paris, UNESCO. http://www.unwater.org/publications/world-water-development-report-2018/

watercalculator (2018): The Hidden Water in Everyday Products – Water Footprint Calculator. http://watercalculator.org

090   Adler, C. et al. (2019): Climate change in the mountain cryosphere. Impacts and responses. Regional Environmental Change. (2019) 19:1225–1228. https://doi.org/10.1007/s10113-019-01507-6

Armstrong, R. L. et al. (2019): Runoff from glacier ice and seasonal snow in High Asia. Separating melt water sources in river flow. https://link.springer.com/article/10.1007%2Fs10113-018-1429-0#Sec14

Ayala, A. et al. (2019): Glacier runoff variations since 1955 in the Maipo River Basin, semi-arid Andes of central Chile. The Cryosphere. https://doi.org/10.5194/tc-2019-233

Cashman, K. (2020): Water Crisis – Parched and Privatised. Development and Cooperation (D+C). https://www.dandc.eu/en/article/chile-faces-serious-water-shortages-due-climate-crisis

The International Commission for the Hydrology of the Rhine basin (CHR) (2018): Disappearing glaciers - New study on melt-water in the Rhine. https://www.chr-khr.org/en/news/disappearing-glaciers-new-study-meltwater-rhine

Pritchard, H. D. (2019): Asia's shrinking glaciers protect large populations from drought stress. Nature volume 569, p. 649–654 (2019). https://www.nature.com/articles/s41586-019-1240-1

Raoul, K. (2014): Can glacial retreat lead to migration? A critical discussion of the impact of glacier shrinkage upon population mobility in the Bolivian Andes. https://link.springer.com/article/10.1007/s11111-014-0226-z

US Bureau of Reclamation (USBR) (2013): Colorado River Basin Water Supply and Demand Study. https://www.usbr.gov/lc/region/programs/crbstudy/FactSheet_June2013.pdf

091

Fountain, A. G. et al. (2018): The Geography of Glaciers and Perennial Snowfields in the American West. Pages 391-410, Published online: 19 Jan 2018 https://www.tandfonline.com/doi/full/10.1657/AAAR0017-003

National Park Service (NPS) (2018): Climate Change. https://www.nps.gov/romo/learn/nature/climatechange.htm

NPS (2023): Glacier and Perennial Snowfield Research in Rocky Mountain National Park.
https://www.nps.gov/articles/glaciers-and-perennial-snowfields-research-in-rocky-mountain-national-park.htm

McGrath, D. 2019. Glacier and Perennial Snowfield Mass Balance of Rocky Mountain National Park (ROMO): Past, Present, and Future. Final Report to Rocky Mountain National Park for Task Agreement P16AC00826. Colorado State University.

092 | 093   DeOreo, W. B., et al. (2016): Residential End Uses of Water, Version 2. Denver, CO: Water Research Foundation.

Cazcarro, I. et al. (2022): Nations' water footprints and virtual water trade of wood products. Advances in Water Resources, 164.
https://www.sciencedirect.com/science/article/pii/S030917082200063X

Mekonnen, M. M. & Hoekstra, A. Y. (2011): National water footprint accounts. The green, blue and grey water footprint of production and consumption, Value of Water Research Report Series No. 50, UNESCO-IHE.

Mekonnen, M. M. & Hoekstra, A. Y. (2012): A global assessment of the water footprint of farm animal products, Ecosystems, 15(3), p. 401–415.

Schyns, J. F., Booij, M. J. & Hoekstra, A. Y. (2017): The water footprint of wood for lumber, pulp, paper, fuel and firewood, Advances in Water Resources 107, p. 490–501. https://waterfootprint.org/en/resources/waterstat/product-water-footprint-statistics/

094   Boucher J., Friot D. (2017): Primary Microplastics in the Oceans: A Global Evaluation of Sources. Gland, Switzerland: IUCN, 2017.
https://portals.iucn.org/library/sites/library/files/documents/2017-002-En.pdf

Westerbos, M., Dagevos, J. (2022): Plastic: The Hidden Beauty Ingredient. Amsterdam: Plastic Soup Foundation.
https://www.beatthemicrobead.org

Centre for Science and Environment (CSE) (2013): 7th State of India's Environment Report: Excreta Matters.
http://cseindia.org/content/excreta-matters-0

United Nations Environment Programme (UNEP) (2021): Drowning in Plastics – Marine Litter and Plastic Waste.
https://www.unep.org/resources/report/drowning-plastics-marine-litter-and-plastic-waste-vital-graphics

095   Bexfield, L. M., et al. (2021): Pesticides and pesticide degradates in groundwater used for public supply across the United States: Occurrence and human-health context. Environmental Science & Technology 55, no.1 (2021): 362–372.
https://doi.org/10.1021/acs.est.0c05793.

DeSimone, L. A., et al. (2015): Water quality in Principal Aquifers of the United States, 1991–2010. Washington, DC: U.S. Geological Survey, Survey Circular 1360, 2015. https://dx.doi.org/10.3133/cir1360.

U.S. Geological Survey (USGS) (2019): Groundwater Quality—Current Conditions and Changes Through Time.
Washington, DC: U.S. Geological Survey, Water Resources Mission Area.
https://www.usgs.gov/mission-areas/water-resources/science/groundwater-quality-current-conditions-and-changes-through.

096 | 097   Circle of Blue (2010): Experts Name the Top 19 Solutions to the Global Freshwater Crisis.
https://www.circleofblue.org/2010/world/experts-name-the-top-19-solutions-to-the-global-freshwater-crisis/

Intergovernmental Panel on Climate Change IPCC (2019): Global Warming of 1,5 °C. Summary for Policymakers.
https://report.ipcc.ch/sr15/pdf/sr15_spm_final.pdf

United Nations (UN) (2021): UN expert: Water crisis is worsening, urgent response needed. News and Press Release.
https://reliefweb.int/report/world/un-expert-water-crisis-worsening-urgent-response-needed

100 | 101   Hipp, S. (Hrsg.) (2020): Kleiner Kompass für mehr Bodenleben.

UBA (2020): Biologische Vielfalt im Boden schützen. https://www.umweltbundesamt.de/biologische-vielfalt-im-boden-schuetzen
UNEP (2015): Soils & Biodiversity. https://i.pinimg.com/736x/29/bc/20/29bc204b52eff70ab2ea66225673af60.jpg

Beiler, K. J. et al. (2010): Architecture of the wood-wide web. Rhizopogon spp. genets link multiple Douglas-fir cohorts.
102 | 103   The New phytologist 185 2 (2010): p. 543–553. DOI: 10.1111/j.1469-8137.2009.03069.x

References

Gorzelak, M. A. et al. (2015): Inter-plant communication through mycorrhizal networks mediates complex adaptive behaviour in plant communities. Oxford University Press / Annals of Botany Company.

Steidinger, B. S. et al. (2019): Climatic controls of decomposition drive the global biogeography of forest-tree symbioses. Nature, 569(7756): p. 404. DOI: 10.1038/s41586-019-1128-0.

104 | 105 Chemnitz, C., Weigelt, J. (2015): Bodenatlas 2015 – Daten und Fakten über Acker, Land und Erde. Heinrich-Böll-Stiftung, Institute for Advanced Sustainability Studies, Bund für Umwelt- und Naturschutz Deutschland und Le Monde diplomatique. https://www.boell.de/de/bodenatlas

Bazyli, C., Kryszak, Ł. (2018): Impact of different models of agriculture on greenhouse gases (GHG) emissions: A sectoral approach. https://doi.org/10.1177/0030727018759092

Flessa, H. et al. (2018): Humus in landwirtschaftlich genutzten Böden Deutschlands. Bundesministerium für Ernährung und Landwirtschaft (BMEL) und Thünen-Institut für Agrarklimaschutz. https://www.bmel.de/SharedDocs/Downloads/Broschueren/Bodenzustandserhebung.pdf?__blob=publicationFile
Mukherjee, A., Kapoor, A. (2018): Organic farming: Learning from Sikkim's experiences. Indian Council for Research on International Economic Relations (ICRIER). https://qrius.com/going-organic-learning-from-sikkims-experiences/

Patzel, N., Wilhelm, B. (2018): Das Boden-Bulletin. Landbau in Zeiten der Erderhitzung, WWF Deutschland.

Wiegandt, K. (Hrsg.) (2022): 3 Grad mehr – Ein Blick in die drohende Heißzeit und wie uns die Natur helfen kann, sie zu verhindern, oekom verlag.

106 | 107 Bai, Y. & Cotrufo, M. F. (2022): Grassland soil carbon sequesttration: Current understanding, challenges and solutions. Science Magazin, 377. https://www.science.org/doi/10.1126/science.abo2380

Buisson, E. et al. (2022): Ancient grasslands guide ambitious goals in grassland restoration. Science Magazin, 377. https://www.science.org/doi/10.1126/science.abo4605

Hipp, S. (Hrsg.) (2020): Kleiner Kompass für mehr Bodenleben.

Isenberg, A. C. (2000): The Destruction of the Bison: An Environmental History , 1750-1920. Cambridge University Press.

108 | 109 Alfred-Wegener-Institut (AWI) (2015): Fact sheet. Das aktuelle Wissen zum Thema: Permafrost. https://www.AWI.de/im-fokus/permafrost.html

Dobiński, W. (2020): Permafrost active layer. Earth-Science Reviews Volume 208, September 2020, 10330. https://www.sciencedirect.com/science/article/abs/pii/S0012825220303470

Ehlers, J. (2011): Eis im Boden – die Formung der Periglazialgebieten: Das Eiszeitalter. Spektrum Akademischer Verlag. https://doi.org/10.1007/978-3-8274-2327-6_8

GRID-Arendal (2020): Coastal and Offshore Permafrost in a Changing Arctic. In: Rapid Response Assessment of Coastal and Offshore Permafrost, Story Map. https://storymaps.arcgis.com/stories/c163de04de7849cdb917fee88015dd73

Overduin, P. P. et al. (2019): Submarine Permafrost Map in the Arctic Modeled Using 1-D Transient Heat Flux (SuPerMAP) in: AGU, JGR Oceans 124, Issue 6. p. 3490-3507. https://doi.org/10.1029/2018JC014675

Obu, J. et al. (2019): Northern Hemisphere permafrost map based on TTOP modelling for 2000–2016 at 1 km^2 scale, in: Earth-Science Reviews 193. p. 299–316. https://doi.org/10.1016/j.earscirev.2019.04.023

110 | 111 Antonelli, A. et al. (2020): State of the World's Plants and Fungi 2020. Royal Botanic Gardens, Kew. DOI: https://doi.org/10.34885/172

Biology Dictionary (BD) (2017): Plants. https://biologydictionary.net/plant/
Landesbetrieb Wald und Holz Nordrhein-Westfalen (2022): Waldweg Grenzenlos. Eichenbaum. https://www.wald-und-holz
.nrw.de/fileadmin/Wald-erleben/Waldweg-Grenzenlos/Manuskript_34_Eichenbaum.pdf

Spektrum (2001): Kompaktlexikon der Biologie. Chloroplasten. https://www.spektrum.de/lexikon/biologie-kompakt/chloroplasten/2352

112 | 113   California Invasive Plant Council (CAL-IPC) (2019): IPCW Plant Report – Eucalyptus globulus.
https://www.cal-ipc.org/resources/library/publications/ipcw/report48/

Pawlik, L. et al. (2016): Roots, Rock, and Regolith. Biomechanical and Biochemical Weathering by Trees and its Impact on Hillslopes. https://www.researchgate.net/profile/ukasz_Pawlik/

Rahul, J. et al. (2015): Adansonia digitata L. (Baobab). A review of traditional information and taxonomic description. Asian Pac J Trop Biomed 2015: 5(1): 79-84. https://www.sciencedirect.com/science/article/pii/S222116911530174X

114 | 115   Beldin, S. I. & Perakis, S. S. (2009): Unearthing the secrets of the forest: U.S. Geological Survey Fact Sheet 2009-3078.
https://pubs.usgs.gov/fs/2009/3078/

Campbell, N. A. et al. (2003): Biologie. Spektrum Akademischer Verlag.

Klein, D. & Schulz, C. (2011): Kohlenstoffspeicherung von Bäumen. LWF-Merkblatt Nr. 27.
http://www.lwf.bayern.de/service/publikationen/lwf_merkblatt/022680/index.php

Klein, D. & Schulz, C. (2011): Der kräftige Atem der Waldböden. LWF aktuell 82, p. 23-25.
http://www.lwf.bayern.de/boden-klima/umweltmonitoring/014491/index.php

Leitgeb, E. (2015): Waldböden und deren nachhaltige Nutzung. BFW-Praxisinformation 39, p. 3-7.
https://www.waldwissen.net/wald/boden/bfw_waldboden_nachhaltig/index_DE#2

Schutzgemeinschaft Deutscher Wald (SDW) (2019): Aufbau des Baumstammes.
http://www.sdw-nrw.de/waldwissen/oekosystem-wald/stammaufbau/

116 | 117   Food and Agriculture Organization of the United Nations (FAO) (2020): Global Forest Resources Assessment 2020. Main report.
Rom. https://doi.org/10.4060/ca9825en

Gauthier, S. et al. (2015): Boreal forest health and global change. Science, 349, p. 819-822. DOI: 10.1126/science.aaa9092.
https://science.sciencemag.org/content/349/6250/819

Jenkins, M. & Schaap, B. (2018): Forest Ecosystem Services – Background Analytical Study 1.
https://www.un.org/esa/forests/wp-content/uploads/2018/05/UNFF13_BkgdStudy_ForestsEcoServices.pdf

Martone, M. et al. (2017): The global forest/non-forest map from TanDEM-X interferometric SAR data.
https://doi.org/10.1016/j.rse.2017.12.002

Petersen, R. et al. (2016): Mapping Tree Plantations with Multispectral Imagery. Preliminary Results for Seven Tropical Countries, World Resources Institute.
https://www.wri.org/research/mapping-tree-plantations-multispectral-imagery-preliminary-results-seven-tropical

Tyrrell, M. L. (2012): Carbon dynamics in the temperate forest, in: Managing Forest Carbon in a Changing Climate, Springer Science & Business Media, p. 77-107. https://link.springer.com/chapter/10.1007/978-94-007-2232-3_5

118 | 119   Critical Ecosystem Partnership Fund (CEPF) (2022): Biodiversity Hotspots Defined.
https://www.cepf.net/our-work/biodiversity-hotspots/hotspots-defined

Conservation International (CI) (2022): Biodiversity Hotspots – Targeted investment in nature's most important places. https://www.conservation.org/priorities/biodiversity-hotspots

Koenig, K. (2016): Biodiversity Hotspots Map (2016.1). Zenodo. https://doi.org/10.5281/zenodo.4311850

120 | 121   Branch, G. (2018): Living Shores – Interacting with Southern Africa's marine ecosystems. Penguin Random House South Africa.

Calloway, M. (2018): Puget Sound Kelp – Trends, Roles and Stressors. Puget Sound Kelp Conservation & Recovery Plan.
https://www.nwstraits.org/media/2802/calloway_kelproletrends6-13-19.pdf

Christie, H. et al. (2003): Species distribution and habitat exploitation of fauna associated with kelp (Laminaria hyperborea) along the Norwegian coast. J. Mar. Biol. Ass. U.K. (2003), 83, 4181/1/13.

References

Estes, J. A. et al. (1998): Killer whale predation on sea otters linking oceanic and nearshore ecosystems. Science 282(5388), p. 473–476. DOI: 10.1126/science.282.5388.473

Filbee-Dexter, K.; Wernberg, T. (2018): Rise of Turfs: A New Battlefront for Globally Declining Kelp Forests. https://academic.oup.com/bioscience/article/68/2/64/4797262

Grebe, G. et al. (2019): An ecosystem approach to kelp aquaculture in the Americas and Europe. https://doi.org/10.1016/j.aqrep.2019.100215

Krause-Jensen, D., Duarte, S. M. (2016): Substantial role of macroalgae in marine carbon sequestration. nature geoscience 9, S. 737–742. https://www.nature.com/articles/ngeo2790

National Oceanic and Atmospheric Administration (NOAA) (2020): Kelp Forests - a Description. Office of National Marine Sanctuaries. https://sanctuaries.noaa.gov/visit/ecosystems/kelpdesc.html
Schiel, D. R.; Foster, M. S. (2016): The Biology and Ecology of Giant Kelp Forests, University of California Press.

Wernberg, T. et al. (2019): Status and Trends for the World's Kelp Forests. World Seas: an Environmental Evaluation, Volume III: Ecological Issues and Environmental Impacts 2019. https://doi.org/10.1016/B978-0-12-805052-1.00003-6

122 | 123   Artaxo, P. et al. (2018): Five Reasons The Earth's Climate Depends on Forests. https://www.climateandlandusealliance.org/scientists-statement/

Centritto, M. et al. (2011): Above Ground Processes. Anticipating Climate Change Influences. https://www.researchgate.net/publication/226698604_Above_Ground_Processes_Anticipating_Climate_Change_Influences

Sanderson, M. et al. (2012): Relationships between forests and weather. EC Directorate General of the Environment. Met Office, Hadley Centre, UK. https://ec.europa.eu/environment/forests/pdf/EU_Forests_annex1.pdf

124 | 125   Betts, R. et al. (2008): Forests and Emissions. A Contribution to the Eliasch Review. Met Office Hadley Centre, UK. https://www.researchgate.net/publication/252653654_Forests_and_Emissions_A_Contribution_to_the_Eliasch_Review

Food and Agriculture Organization of the United Nations (FAO) (2018): The State of the World's Forests – Forest Pathways to Sustainable Development. http://www.fao.org/3/ca0188en/ca0188en.pdf

Harris, N. et al. (2020): Forests Absorb Twice As Much Carbon As They Emit Each Year. https://www.wri.org/insights/forests-absorb-twice-much-carbon-they-emit-each-year

IPCC (2014): Synthesis Report, Fifth Assessment Report. https://www.ipcc.ch/report/ar5/syr/

Max-Planck-Gesellschaft (MPG) (2010): Der Kohlenstoffkreislauf im Erdsystem. https://www.mpg.de/21324/Kohlenstoffkreislauf_im_Erdsystem

Gough, C. M. et al. (2008): Controls on Annual Forest Carbon Storage: Lessons from the Past and Predictions for the Future, in: Bioscience 58(7), p. 609–622. https://academic.oup.com/bioscience/article/58/7/609/237048

World Resources Institute (WRI) (2018): Global Forest Watch. https://www.globalforestwatch.org

126 | 127   Great Green Wall (GGW) (2020): Growing a world wonder. https://www.greatgreenwall.org

UNCCD (2020): The Great Green Wall. Implementation Status and Way Ahead to 2030. https://www.unccd.int/resources/publications/great-green-wall-implementation-status-way-ahead-2030

128 | 129   BOM (2019): Special Climate Statement 70 update—drought conditions in Australia and impact on water resources in the Murray–Darling Basin. http://www.bom.gov.au/climate/current/statements/scs70.pdf

Fei, S. et al. (2017): Divergence of species responses to climate change. Science Advances 3(5). DOI: 10.1126/sciadv.1603055. https://advances.sciencemag.org/content/3/5/e1603055/tab-pdf

Forsyth, G. G. et al. (2004): A rapid assessment of the invasive status of Eucalyptus species in two South African provinces. https://open.uct.ac.za/bitstream/item/17766/Forsyth_A_rapid_assessment_2004.pdf?sequence=1

Kolb, P. (2019): Tree-Climate Interactions. United States Department of Agriculture.
https://climate-woodlands.extension.org/treeclimate-interactions/

Malhi, Y., Philips, O. L. (2005): Tropical Forests and Global Atmospheric Change.
https://www.researchgate.net/publication/8496341_Tropical_Forests_and_Global_Atmospheric_Change

Mukherjee, S. et al. (2022): Hotspots shed light on »flash drought« causes. National Science Foundation.
https://phys.org/news/2022-02-hotspots-drought.html

Nature Climate Change (NCC) (2020): In the line of fire. Editorial. Nature Climate Change 10(169).
https://doi.org/10.1038/s41558-020-0720-5

Vose, J. M. et al. (2015): Effects of Drought on Forests and Rangelands in the United States. A Comprehensive Science Synthesis.
Forest Service Gen. Tech. Report. https://www.fs.usda.gov/sites/default/files/GTR-WO-93a.pdf

130 | 131   Dinerstein, E. (2020): A »Global Safety Net« to reverse biodiversity loss and stabilize Earth's climate.
https://www.science.org/doi/10.1126/sciadv.abb2824

132 | 133   IUCN (2022): IUCN Red List. https://www.iucn.org

Mora, C. et al. (2011): How many species are there on Earth and in the ocean? PLoS Biology, 9(8), e1001127.

Wiegandt, K. (Hrsg.) (2022): 3 Grad mehr. Ein Blick in die drohende Heißzeit und wie uns die Natur helfen kann, sie zu
verhindern, oekom verlag.

134 | 135   Eulen- und Greifvögelstation Haringsee (2021): Wissenswertes über Eulen und Greifvögel.
https://www.eulen-greifvogelstation.at/wissen-ueber-wildtiere/wissenswertes-ueber-eulen-und-greifvoegel

Euronatur: Steckbrief Braunbär. https://www.euronatur.org/unsere-themen/artenschutz/braunbaer/steckbrief-baer/

Froschnetz: Frösche, Kröten, Molche. https://www.froschnetz.ch

Jedrzejewski, W.; Jedrzejewski, B. (1992): Foraging and diet of the red fox Vulpes vulpes in relation to variable food
resources in Biatowieza National Park, Poland, in: Ecography 15(2), p. 212-220.

Luchsprojekt Bayern: Nahrung. http://www.luchsprojekt.de/06_lebensweise/nahrung.html

Schlangen in Deutschland: 6 Heimische Schlangen leben bei uns. https://www.schlangen-in-deutschland.de

NABU: Der Buntspecht. Vogel des Jahres 1997.
https://www.nabu.de/tiere-und-pflanzen/aktionen-und-projekte/vogel-desjahres/1997-buntspecht

Neuschulz, E. L. et al. (2016): Pollination and seed dispersal are the most threatened processes of plant regeneration.
Nature. Scientific Reports. http://www.nature.com/articles/srep29839

Niethammer, J. & Krapp, F. (1986): Handbuch der Säugetiere Europas. Band 2/II Paarhufer – Artiodactyla. Aula Verlag.

Wildtiermanagement Niedersachsen: Baummarder (Martes martes).
https://www.wildtiermanagement.com/wildtiere/haarwild/baummarder

Wolfsinformationszentrum Schleswig-Holstein: Biologie des Wolfes. https://www.wolfsinfozentrum.de/biologie-des-wolfes-1.html

136 | 137   Alroy, J. (2017): Effects of habitat disturbance on tropical forest biodiversity. 6056–6061 PNAS, 114(23).
https://www.pnas.org/content/pnas/114/23/6056.full.pdf

Internationel Union for Conservation of Nature (IUCN) (2020): The IUCN Red list of threatened Species.
https://www.iucnredlist.org

138 | 139   Bassett, Y., Lamarre, G. P. A. (2019): Toward a world that values insects. Science Magazin, 364(6447), S. 1230-1231.
https://www.science.org/doi/10.1126/science.aaw7071

References

Eggleton, P. et al. (2020): The State of the World's Insects.
https://www.annualreviews.org/doi/pdf/10.1146/annurev-environ-012420-050035

Kleijn, D. et al. (2016): Delivery of crop pollination services is an insufficient argument for wild pollinator conservation. Nature Communications. https://www.nature.com/articles/ncomms8414.pdf

Sánchez-Bayo, F., Wyckhuys, K. A. (2019): Worldwide decline of the entomofauna: A review of its drivers. Biological Conservation 232 (2019): 8–27. https://doi.org/10.1016/j.biocon.2019.01.020.

Seibold, S. S. et al. (2019): Arthropod decline in grasslands and forests is associated with landscape-level drivers. Nature, 574(7780), p. 671–674. https://pubmed.ncbi.nlm.nih.gov/31666721/

United States Department of Agriculture (USDA) (2022): The Importance of Pollinators.
https://www.usda.gov/peoples-garden/pollinators

140 | 141   Center for Biological Diversity (CBD) (2021): Emperor Penguin. Natural History.
https://www.biologicaldiversity.org/species/birds/penguins/emperor_penguin.html

Davidson, S. C. et al. (2020): Ecological insights from three decades of animal movement tracking across a changing Arctic. Science, 370(6517), p. 712–715. https://www.science.org/doi/10.1126/science.abb7080

Drost, H. E. et al. (2016): Acclimation potential of Arctic cod (Boreogadus saida) from the rapidly warming Arctic Ocean. https://jeb.biologists.org/content/jexbio/219/19/3114.full.pdf

Jenouvrier, S. et al. (2014): Projected continent-wide declines of the emperor penguin under climate change. Nature Climate Change, 4, p. 715–718. https://www.nature.com/articles/nclimate2280

Klein, D. R., Sowls, A. (2015): Red Foxes Replace Arctic Foxes on a Bering Sea Island. Consequences for Nesting Birds.
https://www.nps.gov/articles/aps-v14-i1-c5.htm

Lefort, K. J. et al. (2020): Killer whale abundance and predicted narwhal consumption in the Canadian Arctic.
https://doi.org/10.1111/gcb.15152

Thyrrin, J., Sejr, M. K. (2018): Climate Change draws invasive species to the Arctic. Science Nordic. https://sciencenordic.com/animals--plants-climatechange-denmark/climate-change-draws-invasive-species-to-the-arctic/1454214

142 | 143   Barbet-Massin, M. et al. (2019): The economic cost of control of the invasive yellow-legged Asian hornet. NeoBiota 55, p. 11–25. http://neobiota.pensoft.net

Bellard, C. et al. (2013): Will climate change promote future invasions? Global Change Biology. https://doi.org/10.1111/gcb.12344

Ling et al. (2009): Overfishing reduces resilience of kelp beds to climate-driven catastrophic phase shift. Proceedings of the National Academy of Sciences 106(52), p. 22341-22345. https://www.researchgate.net/publication/40696070_Overfishing_reduces_resilience_of_kelp_beds_to_climate-driven_catastrophic_phase_shift

Lockwood, J. L. et al. (2019): When pets become pests. The role of the exotic pet trade in producing invasive vertebrate animals. Frontiers in Ecology and the Environment. https://doi.org/10.1002/fee.2059

Vickery, P. D. et al. (2020): Birds of Maine. Princeton University Press.

144 | 145   Grosberg, R. K. et al. (2012): Biodiversity in water and on land. Current Biology, 22(21). https://doi.org/10.1016/j.cub.2012.09.050

Maribus (2010): World Ocean Review. Mit den Meeren leben. http://worldoceanreview.com

Mittermeier, R. A. et al. (2011): Global biodiversity conservation. The critical role of hotspots. Springer.
Poulsen, J. Y. et al. (2016): Preservation Obscures Pelagic Deep-Sea Fish Diversity. Doubling the Number of Sole-Bearing Opisthoproctids and Resurrection of the Genus Monacoa (Opisthoproctidae, Argentiniformes). PLoS ONE 11(8): e0159762.

146 | 147   Abdulla, A. et al. (2013): Marine Natural Heritage and the World Heritage List. Interpretation of World Heritage criteria in marine systems, analysis of biogeographic representation of sites, and a roadmap for addressing gaps. IUCN.

IUCN (2017): The IUCN Red List of Threatened Species. http://www.iucnredlist.org

148 | 149   EU Fishing Fleet Register (2016): Fishing fleet capacities.
https://oceans-and-fisheries.ec.europa.eu/fisheries/rules/fishing-fleet-capacities_en

Greenpeace (2014): Fish Fairly. http://www.greenpeace.de/fairfischen

Reedereien (2017): Havfisk, Norway. http://www.havfisk.no/en
Parlevliet en Van der Plas B.V.: Fish for Life. http://parlevliet-vanderplas.nl/

150 | 151   Global Fishing Watch (GFW) (2021): Apparent Fishing Effort. https://globalfishingwatch.org/map

Watson, R. et al. (2012): Spatial expansion of EU and non-EU fishing fleets into the global ocean 1950 to present. The Sea around
us-Project, Fish Centre Univ. British Columbia, Kanada, und World Wildlife Fund (WWF)
http://wwf.panda.org/wwf_news/?203247/Wild-west-fishing-in-distant-waters

152 | 153   Department of Fisheries and Oceans Canada (DFO) (2013): Aquaculture in Canada: Integrated Multi-Trophic Aquaculture (IMTA)
http://publications.gc.ca/collections/collection_2013/mpo-dfo/Fs45-4-2013-eng.pdf

Maribus (2013): World Ocean Review 2. Die Zukuft der Fische, die Fischerei der Zukunft.

154 | 155   Duarte, C. M. et al. (2021): The Soundscape of the Anthropocene ocean. Science, 371, eaba4658.
https://science.sciencemag.org/content/371/6529/eaba4658.full

Nowacek, S. M. et al. (2001): Short-term effects of boat traffic on bottlenose dolphins, tursiops truncatus,
in Sarasota Bay, Florida. Marine Mammal Science, 17(4), p. 673-688.

Schorr, G. S. et al. (2014): First Long-Term Behavioral Records from Cuvier's Beaked Whales
(Ziphiuscavirostris) Reveal Record-Breaking Dives. PLoS One. 9(3): e92633. DOI: 10.1371/journal.pone.0092633

Veirs, S. et al. (2016): Ship noise extends to frequencies used for echolocation by endangered killer whales.
PeerJ 4:e1657. https://doi.org/10.7717/peerj.1657

156 | 157   Australian Government Department of the Environment (AGDE) (2016): Commonwealth Marine Reserves Review. Goals and Prin-
ciples. http://www.environment.gov.au/marinereservesreview/goals-principles

UN Environment Programm–World Conservation Monitoring Center (UNEP-WCMC), International Union for Conservation of
Nature (IUCN) (2021): Protected Planet Report 2020. https://livereport.protectedplanet.net/

Sciberras, M. et al. (2015): Evaluating the relative conservation value of fully and partially protected marine areas.
Fish and Fisheries, 16, p. 58-77. DOI: 10.1111/faf.12044

160 | 161   Vienna University of Economics and Business (WU Wien) (2022): Material flows by material group, 1970-2019. UN-IRP Global
Material Flows Database. https://www.materialflows.net/visualisation-centre

WU Wien (2022): Fineprint Geovisualisations. https://www.fineprint.global/resources/mining-areas/

162 | 163   GEOMAR (2016): Massivsulfide. Rohstoffe aus der Tiefsee.
http://www.geomar.de/fileadmin/content/service/presse/public-pubs/massivsulfide_2016_de_web.pdf

International Energy Agency (IEA) (2018): Offshore Energy Outlook 2018. World Energy Outlook Special Report.
https://www.iea.org/reports/offshore-energy-outlook-2018

Janssen, F. et al. (2020): Manganknollen-Abbau gefährdet die Ökosysteme der Tiefsee. Potsdam Earth System Knowledge Platform
(ESKP). https://doi.org/10.2312/eskp.023

Maribus (2014): World Ocean Review 3. Rohstoffe aus dem Meer. Chancen und Risiken.
http://worldoceanreview.com

Rona, P. A. (2003): Resources of the sea floor. Science. http://science.sciencemag.org/content/299/5607/673

References

Umweltbundesamt (UBA) (2013): Tiefseebergbau und andere Nutzungsarten der Tiefsee. http://www.umweltbundesamt.de/themen/wasser/gewaesser/meere/nutzung-belastungen/tiefseebergbau-andere-nutzungsarten-der-tiefsee

United Nations Environmental Programme (UNEP) (2013): Wealth in the Oceans: Deep sea mining on the horizon? UNEP Global Environmental Alter Service. http://www.unep.org/geas

164 | 165   Centre of Documentation, Research and Experimentation on Accidental Water Pollution (CEDRE) (2016): Database of spill incidents and threats in waters around the world. http://wwz.cedre.fr/en/Our-resources/Spills

Maribus (2010): World Ocean Review 1. Mit den Meeren leben. http://worldoceanreview.com

National Academies of Sciences, Engineering, and Medicine (NA) (2022): Oil in the Sea IV. Inputs, Fates, and Effects. The National Academies Press. https://doi.org/10.17226/26410

Safety4sea (2020): 1978-2020: List of major oil spills. https://safety4sea.com/1978-2020-list-of-major-oil-spills/

Woods Hole Oceanic Institution (WHOI) (2011): Oil in the Ocean. A Complex Mix. https://www.whoi.edu/oilinocean/page.do?pid=51878

166 | 167   U.S. Energy Information Administration (EIA) (2024): What is U.S. electricity generation by energy source? Electric Power Monthly, 2024. https://www.eia.gov/electricity/monthly

U.S. Census Bureau (USCB) (2022): American Community Survey, 2022 American Community Survey 1-Year Estimates, Table B25040. https://data.census.gov

World Integrated Trade Solution (WITS) (2021): United States Fuel Imports: by country and region in US $ 2021. https://wits.worldbank.org/CountryProfile/en/Country/USA/Year/2021/TradeFlow/Import/Partner/ALL/Product/Fuels

168 | 169   Bundesministerium für Umwelt, Naturschutz, nukleare Sicherheit und Verbraucherschutz (BMUV) (2022): Fragen und Antworten zur AKW-Laufzeitverlängerung. https://www.bmuv.de/themen/atomenergie-strahlenschutz/nukleare-sicherheit/faq-akw-laufzeitverlaengerung

Centers for Disease Control and Prevention (CDC) (2020): Worker Health Study Summaries. Uranium Miners. https://www.cdc.gov/niosh/pgms/worknotify/uranium.html

EIA (2022): Nuclear explained. U.S. nuclear industry. https://www.eia.gov/energyexplained/nuclear/us-nuclear-industry.php

EIA (2022): Annual Energy Outlook 2022. https://www.eia.gov/outlooks/aeo/narrative/electricity/sub-topic-02.php

World Nuclear Association (WNA) (2022): World Uranium Mining Production. https://world-nuclear.org/information-library/nuclear-fuel-cycle/mining-of-uranium/world-uranium-mining-production.aspx

WNA (2022): Nuclear Power in France. https://world-nuclear.org/information-library/country-profiles/countries-a-f/france.aspx

170 | 171   Food and Agriculture Organization of the United Nations (FAO) (2018): The State of the World's Forests. Forest Pathways to Sustainable Development. http://www.fao.org/3/ca0188en/ca0188en.pdf

FAO (2020): Forestry Production and Trade. http://www.fao.org/faostat/en/#data/FO

Goncalves, A. C. et al. (2018): Solid Biomass from Forest Trees to Energy. A Review. Intech Open. https://www.intechopen.com/books/renewable-resources-and-biorefineries/solid-biomass-from-forest-trees-to-energy-a-review

Sterman, J. D. et al. (2018): Does replacing coal with wood lower CO2 emissions? Dynamic lifecycle analysis of wood bioenergy Environmantal Research Letter. 13. https://iopscience.iop.org/article/10.1088/1748-9326/aaa512/pdf

U.S. International Trade Commission (USITC) (2018): International Trade in Wood Pellets: Current Trends and Future Prospects. https://www.usitc.gov/publications/332/executive_briefings/wood_pellets_ebot_final.pdf

Voegele, E. (2022): US pellet exports reach 7.52 million metric tons in 2021. United States Department of Agriculture. https://biomassmagazine.com/articles/18705/usda-us-pellet-exports-reach-7-52-million-metric-tons-in-2021

World Bioenergy Association (WBA) (2017): Global Bioenergy Statistics 2017.
https://worldbioenergy.org/uploads/WBA%20GBS%202017_hq.pdf

172 | 173   Schmidt-Curelli, J. & Knebel, A. (2017): Energiewendeatlas Deutschland 2030. Agentur für Erneuerbare Energien e. V.
https://www.unendlich-viel-energie.de/mediathek/publikationen/energiewendeatlas-deutschland-2030

Jacobson, M. Z. et al. (2017): 100% Clean and Renewable Wind, Water, and Sunlight: All-Sector Energy Roadmaps for 139 Countries of the World. Joule 1, no.1 (2017): 108–121. http://dx.doi.org/10.1016/j.joule.2017.07.005

174   McGinley, K., et al. (2023): National Report on Sustainable Forests. Washington, DC: U.S. Department of Agriculture.
https://www.fs.usda.gov/sites/default/files/fs_media/fs_document/2020-sustainability-report.pdf.

U.S. Forest Service (USFS) (2022): Forest Products. U.S. Department of Agriculture. https://www.fs.usda.gov/research/forestproducts.

175   Food and Agriculture Organization of the United Nations (FAO) (2019): Global Forest Products Facts and Figures 2018.
http://www.fao.org/3/ca7415en/ca7415en.pdf

FAO (2020): Forestry Production and Trade. http://www.fao.org/faostat/en/#data/FO

Hetemäki, L. & Hurmekoski, E. (2016): Forest Products Markets under Change: Review and Research Implications.
https://link.springer.com/article/10.1007/s40725-016-0042-z

Internationale Energie Agentur (IEA) (2022): Pulp and paper. Not on track. Tracking Report. September 2022.
https://www.iea.org/reports/pulp-and-paper

UNECE/FAO (2019): Forest Products. Annual Market Review 2018–2019. United Nations Publication.
http://www.unece.org/fileadmin/DAM/timber/publications/SP48.pdf

176 | 177   Dewi, I. K. et al. (2019): Implementation of environmental management policies on the impact of illegal sand mining . IOP Conf. Ser.: Earth Environ. Sci. 343 012129. https://iopscience.iop.org/article/10.1088/1755-1315/343/1/012129/pdf

Langrand, M., Peduzzi, P. (2022): UN environment: world needs to ditch sand addiction. Geneva Solutions. Climate & Environment Interview. https://genevasolutions.news/climate-environment/un-environment-world-needs-to-ditch-sand-addiction

Masalu, D. C. P. (2019): Coastal Erosion and Its Social and Environmental Aspects in Tanzania: A Case Study in Illegal Sand Mining Coastal Management 30, p. 347-359. https://www.semanticscholar.org/paper/Coastal-Erosion-and-Its-Social-and-Environmental-in-Masalu/ac8c1d27aa51ae40887c1a9430841dd43ab5cdd5

UNEP (2022): Sand and Sustainability: 10 Strategic Recommendations to Avert a Crisis.
https://www.unep.org/resources/report/sand-and-sustainability-10-strategic-recommendations-avert-crisis

178 | 179   Schlining, K. et al. (2013): Debris in the deep: Using a 22-year video annotation database to survey marine litter in Monterey Canyon, Central California, USA. Monterey Bay Aquarium Research Institute (MBARI).

Petroleum Economist (2019): Plastic recycling threatens oil demand growth. https://www.petroleum-economist.com/articles/mid stream-downstream/refining-marketing/2019/plastic-recycling-threatens-oil-demand-growth

Plastics Europe (2016): EU Plastics Production and Demand First Estimates for 2020.
https://www.plasticseurope.org/en/newsroom/news/eu-plastics-production-and-demand-first-estimates-2020

Subba Reddy, M. et al. (2014): Effect of Plastic Pollution on Environment. Department of Chemistry, S.B.V.R. Aided Degree College, badvel, Kadapa-516227, India. Journal of Chemical and Pharmaceutical Sciences.

United Nations Environment Programme (UNEP) (2015): The Plastics Disclosure Project.
http://www.plasticdisclosure.org/about/why-pdp.html
World Economic Forum (WEF) (2016): The New Plastics Economy. Rethinking the future of plastics.
https://www.weforum.org/reports/the-new-plastics-economy-rethinking-the-future-of-plastics

180 | 181   Europäische Kommission (EC) (2022): Circular economy action plan.
https://environment.ec.europa.eu/strategy/circular-economy-action-plan_de#actions

References

WU (2022): The importance of Circular Economy actions for GHG emission reductions. http://www.materialflows.net/stories

182 | 183   Airports Council International (ACI) (2018): Statistics. World Airport Traffic 2018.
https://aci.aero/news/2018/09/20/aci-world-publishes-annual-world-airport-traffic-report/

European Space Agency (ESA) (2016): Satellite-AIS-based map of global ship traffic. https://www.esa.int/ESA_Multimedia
/Images/2016/01/Satellite-AIS-based_map_of_global_ship_traffic

Greene, S. (2023): Freight Transportation. Massachusetts Institute of Technology, MIT Sustainable Supply Chains Initiative.
https://climate.mit.edu/explainers/freight-transportation

Statista (2024): Container shipping – statistics & facts https://www.statista.com/topics/1367/container-shipping/#topicOverview

184 | 185   Kommission der Europäischen Gemeinschaften, Statistisches Amt (KEG) (1970): Der Seeverkehr der Länder der Gemeinschaft.
1955, 1960 und 1967. Eine statistische Studie, Brüssel-Luxemburg, Mai 1970.

NABU (2014): Luftschadstoffemissionen von Containerschiffen. Hintergrundpapier. https://www.nabu.de/imperia/md/content
/nabude/verkehr/140623-nabu-hintergrundpapier_containerschifftransporte.pdf

United Nations Conference on Trade and Development (UNCTAD) (2020): Review of Maritime Transport 2020. United Nations.
https://unctad.org/system/files/official-document/rmt2020_en.pdf

186 | 187   American Trucking Associations (ATA) (2022): Economics and Industry Data.
https://www.trucking.org/economics-and-industry-data

U.S. Department of Transportation Federal Highway Commission (DOT) (2022): Highway Statistics 2022. Washington, D.C: Policy and
Governmental Affairs, Office of Highway Policy Information. https://www.fhwa.dot.gov/policyinformation/statistics/2022

U.S. Environmental Protection Agency (EPA) (2023): Fast Facts: U.S. Transportation Sector Greenhouse Gas Emissions, 1990-2021."
Washington, DC: Office of Transportation and Air Quality. https://www.epa.gov/system/files/documents/2023-06/420f23016.pdf

Freemark, Y. (2020): Too Little, Too Late? A Decade of Transit Investment in the U.S. Streetsblog USA. January 8, 2020.
https://usa.streetsblog.org/2020/01/08/too-little-too-late-a-decade-of-transit-investment-in-the-u-s.

188 | 189   Bebbington, A. J. et al. (2018): Resource extraction and infrastructure threaten forest cover and community rights.
https://www.pnas.org/content/115/52/13164

Yale School of Forestry and Environmental Studies (2020): Global Forest Atlas. Roads and Forests.
https://globalforestatlas.yale.edu

190 | 191   Livesley, S. J. et al. (2016): The Urban Forest and Ecosystem Services: Impacts on Urban Water, Heat, and Pollution Cycles at the
Tree, Street, and City Scale. J. Environ. Qual. 45:119–124 (2016). doi:10.2134/jeq2015.11.0567

Marx, A. (Hrsg.): Klimaanpassung in Forschung und Politik. Helmholtz-Zentrum für Umweltforschung GmbH, UFZ Leipzig.
Springer Spektrum.

Moser, A. et al. (2017): Stadtbäume. Wachstum, Funktionen, und Leistungen. Risiken und Forschungsperspektiven. Allgemeine
Forst und Jagdzeitung, Jg. 5/6, p. 188ff. http://waldwachstum.wzw.tum.de/fileadmin/publications/Moser_2018.pdf

Rahman, M. A. et al. (2018): Vertical air temperature gradients under the shade of two contrasting urban tree species during
different types of summer days. Science of The Total Environment 633. DOI: 10.1016/j.scitotenv.2018.03.168

Roloff, A. (2013): Bäume in der Stadt. Besonderheiten – Funktion – Nutzen – Arten – Risiken. Ulmer.

Winbourne, J. B. et al. (2020): Tree Transpiration and Urban Temperatures: Current Understanding, Implications, and Future
Research Directions. BioScience, 70, no.7 (July 2020): 576–588. https://doi.org/10.1093/biosci/biaa055.

192 | 193   FAO (2021): Small family farmers produce a third of the world's food. New FAO research focuses on contributions of farmers with
fewer than two hectares. https://www.fao.org/news/story/en/item/1395127/icode/

Fernandez, L. (2022): Global pesticide agricultural use 2020, by leading country. Statista. https://www.statista.com/statistics /1263069/global-pesticide-use-by-country/

Lowder, S. et al. (2021): Which farms feed the world and has farmland become more concentrated? https://www.sciencedirect .com/science/article/pii/S0305750X2100067X?via%3Dihub

OECD-FAO (2022): AGRICULTURAL OUTLOOK 2022–2031. https://www.oecd.org/publications/oecd-fao-agricultural-outlook-19991142.htm

Sharma, A. et al. (2019): Worldwide pesticide usage and its impacts on ecosystem. Review Paper. Springer Nature. https://link.springer.com/content/pdf/10.1007/s42452-019-1485-1.pdf

194 | 195   Gerber, P. J. et al. (2013): Tackling climate change through livestock. A global assessment of emissions and mitigation opportunities. FAO.

Guégan, S., Léger, F. (2015): Case Study. Permacultural Organic Market Gardening and Economic Performance.

IPCC (2019): Climate Change and Land. An IPCC Special Report on climate change, desertification, land degradation, sustainable land management, food security, and greenhouse gas fluxes in terrestrial ecosystems. https://www.ipcc.ch/site/assets/uploads/2019/08/Fullreport-1.pdf

Thornton, P. et al. (2018): Agriculture in a changing climate. Keeping our cool in the face of the hothouse. https://journals.sagepub.com/doi/10.1177/0030727018815332

De Ramon N'Yeurt, A. et al. (2012): Negative Carbon Via Ocean Afforestation. https://www.researchgate.net/publication/259892834_Negative_Carbon_Via_Ocean_Afforestation

Ritchie, H., Roser, M. (2020): Environmental Impacts of Food Production. OurWorldInData.org. https://ourworldindata.org/environmental-impacts-of-food

196 | 197   Franklin, J. F. (2018): Ecological Forest Management. Waveland Press, Inc.

Food and Agriculture Organization of the United Nations (FAO) (2019): Agroforestry. http://www.fao.org/forestry/agroforestry/en/

Deutscher Fachverband für Agroforstwirtschaft (DeFAF) (2020): Arten von Agroforstsystemen. https://agroforst-info.de/arten

198 | 199   Carbon Footprint (2022): Carbon Calculator. Carbon Footprint Calculator For Individuals And Households. https://www.calculator.carbonfootprint.com

200 | 201   KlimAktiv (2020): Klimaschutz mit Handprint. https://www.climate-handprint.de

202 | 203   Captain Planet Foundation (2022): JOHN ABAD. https://captainplanetfoundation.org/about/our-people/john-abad/

Hindustan Times (2022): Meet Saalumarada Thimmakka, 111-year-old environmentalist given a cabinet rank. https://www.hindustantimes.com/cities/bengaluru-news/meet-saalumarada-thimmakka-111-year-old-environmentalist-given -a-cabinet-rank-101657513076203.html

Lehmann, R. M. (2022): 100.000€ in 3h – Der Wendepunkt des Internets | #HaiBockAngriff. Mission Erde. https://www.youtube.com/watch?v=IEjHYRn7840

Smithsonian (2022): Pablo Garcia Borboroglu. https://ocean.si.edu/contributors/pablo-garcia-borboroglu

UNEP (2016): Champions of the Earth. Afroz Shah – Inspiration and Action. https://www.unep.org/championsofearth/laureates/2016/afroz-shah

About the Author

When Esther Gonstalla is not outdoors surfing, hiking, biking or kayaking, she works as a freelance infographic artist and environmental book author. Her clients include NGOs, scientific institutes, and magazines such as *National Geographic* Germany, German Ocean Foundation, Bread for the World (Germany), Friends of the Earth (Germany), Coalition for Fair Fisheries Arrangements (Belgium), Global Water Partnership (Sweden), Institute for Climate Physics (South Korea) and University of Manoa, Hawaii (USA).

Since graduating in Information Design from Münster University of Applied Sciences in 2009, Esther has worked as a freelancer, taking full advantage of the freedom to move around for 10 years, working in a different country every year. Since 2022, Esther has lived with her husband on the Great Ocean Road in Victoria, Australia.

Esther designed and wrote the first book of its kind in 2009 as her university thesis: The Atomic Book: Radioactive Waste and Lost Atomic Bombs. It was awarded the prestigious Artbook Prize (Stiftung Buchkunst) as "one of the most beautiful German books of 2009." Esther created the explainer book series "Our World in 50 Graphics" published by oekom verlag, Munich, Germany in 2017. The series includes The Ocean Book, The Climate Book, The Forest Book, and The Ice Book.

Portfolio: www.gonstalla.com   **Instagram:** @infographic_nomad

Thanks!

This book is a tribute to the decades of work by more than a thousand scientists who have contributed to the hundreds of studies whose findings are presented here.

A huge thank you for the professional support goes to: **Dr. Thorben Amann** (University of Hamburg), **Prof. Dr. Christian Ammer** (University of Göttingen), **Dr. Inka Bartsch** (Alfred-Wegener-Institute), **Hans von der Goltz** (Arbeitsgemeinschaft Naturgemäße Waldwirtschaft Deutschland e. V.), **Prof. Dr. Hartmut Graßl** (Max Planck Institute for Meteorology), **Dr. Inge Grünberg** (AWI), **Dr. Sven Günter** (Thünen-Institute), **Prof. Robert Hänsch** (Braunschweig University of Technology), **Dr. Sabine Henders** (climatekos), **Prof. Pierre Ibisch** (Eberswalde University for Sustainable Development), **Prof. Peter Irvine** (University College London), **Prof. Hans Joosten** (University of Greifswald), **Prof. Dr. Daniela Kleinschmit** (University of Freiburg), **Prof. Dr. Michael Köhl** (University of Hamburg), **Dr. Christoph Mayer** (Bavarian Academy of Sciences and Humanities), **Dr. Marcel Nicolaus** (AWI), **Prof. Dr. Barbara Niehoff** (AWI), **Prof. Andreas Oschlies** (GEOMAR), **Dr. Paul Overduin** (AWI), **Prof. Dr. Daniel Pauly** (University of Vancouver), **Dr. Nikola Patzel** (Bureau for Soil Communication), **Dr. Sven Petersen** (GEOMAR), **Prof. Volker Quaschning** (HTW Berlin), **Dr. Ingo Sasgen** (AWI), **Dr. Fabian Schlösser** (University of Hawaii), **Janine Schmidt-Curelli** (Climate Protection and Energy Agency kleVer), **Prof. Kai Schröter** (TU Braunschweig), **Stefan Schwarzer** (Bureau for Soil Communication Geneva), **Frank Schweikert** (German Ocean Foundation), **Prof. Axel Timmermann** (IBS Center for Climate Physics, South Korea) and **Dr. Susanne Winter** (TU Dresden).

The following graphics are produced in collaboration with ...
P. 38-39: Prof. Axel Timmermann, IBS Center for Climate Physics, Busan, South Korea
P. 154-155: Philippe Pajot, "La Recherche" magazine, Paris, France
P. 176-177: Janine Schmidt-Curelli, Climate Protection and Energy Agency kleVer, Verden (Aller), Germany
... and have been updated or adapted for this book. Thank you very much!

A big thank you goes to my husband, Craig Martin, who helped me translate this atlas and who inspired me to go on adventurous sea kayaking trips in between writing sessions.

Many thanks to my wonderful publishers: Dr. Laura Kohlrausch from oekom verlag for the many years of cooperation and to Rebecca Bright from Island Press for the great job in editing the English version.

About Island Press

Since 1984, the nonprofit organization Island Press has been stimulating, shaping, and communicating ideas that are essential for solving environmental problems worldwide. With more than 1,000 titles in print and some 30 new releases each year, we are the nation's leading publisher on environmental issues. We identify innovative thinkers and emerging trends in the environmental field. We work with world-renowned experts and authors to develop cross-disciplinary solutions to environmental challenges.

Island Press designs and executes educational campaigns, in conjunction with our authors, to communicate their critical messages in print, in person, and online using the latest technologies, innovative programs, and the media. Our goal is to reach targeted audiences—scientists, policy makers, environmental advocates, urban planners, the media, and concerned citizens—with information that can be used to create the framework for long-term ecological health and human well-being.

Island Press gratefully acknowledges major support from The Bobolink Foundation, Caldera Foundation, The Curtis and Edith Munson Foundation, The Forrest C. and Frances H. Lattner Foundation, The JPB Foundation, The Kresge Foundation, The Summit Charitable Foundation, Inc., and many other generous organizations and individuals.

The opinions expressed in this book are those of the author(s) and do not necessarily reflect the views of our supporters.

"Our Earth is changing at an
unprecedented rate.
The time for action has
long since arrived.
Millions of people are working
for conservation,
sustainable lifestyles, and
a livable climate.
Let's all get involved!"

Esther Gonstalla